"十三五"职业教育部委级规划教材
2018江苏省高等学校重点教材 2018-2-125
江苏高校品牌专业建设工程资助项目（PPZY2015C254）

纺织导论双语教程

Bilingual Course in Textile Introduction

赵　磊　陈宏武　主　编
王　可　陈贵翠　副主编

中国纺织出版社有限公司　　国家一级出版社
全国百佳图书出版单位

内 容 提 要

本书在纺织产业转型升级与转移趋势明显的背景下，结合纺织产业对从业人员提出纺织技术技能和英语素养的双重要求，与市场充分接轨，以项目任务为载体，介绍纺织材料、纺纱技术、机织技术、针织技术、非织造技术、纺织品染整技术，内容定位精准明确，言简意赅。

本书可供纺织服装职业教育院校相关专业师生作为教材使用，也可供留学生拓展纺织专业基础知识参考，还可供纺织生产和外贸等从业人员作为双语知识学习资料。

图书在版编目(CIP)数据

纺织导论双语教程/赵磊，陈宏武主编. ——北京：中国纺织出版社有限公司，2019.10（2023.8重印）

"十三五"职业教育部委级规划教材 2018江苏省高等学校重点教材

ISBN 978-7-5180-6624-7

Ⅰ.①纺… Ⅱ.①赵…②陈… Ⅲ.①纺织—双语教学—高等职业教育—教材 Ⅳ.①TS1

中国版本图书馆CIP数据核字（2019）第188009号

责任编辑：符 芬　责任校对：王花妮　责任印制：何 建

中国纺织出版社有限公司出版发行
地址：北京市朝阳区百子湾东里A407号楼　邮政编码：100124
销售电话：010—67004322　传真：010—87155801
http://www.c-textilep.com
E-mail:faxing@c-textilep.com
中国纺织出版社天猫旗舰店
官方微博 http://weibo.com/2119887771
北京虎彩文化传播有限公司印刷　各地新华书店经销
2023年8月第3次印刷
开本：787×1092　1/16　印张：16.25　彩插：0.5
字数：500千字　定价：56.00元

凡购本书，如有缺页、倒页、脱页，由本社图书营销中心调换

前　言

中国是世界上一大纺织服装生产国与出口国。一方面,近年来,国内纺织产业发展重心进一步向"微笑曲线"两端延伸,实现产业升级;另一方面,纺织产业转移趋势明显,向具有产业优势的世界各地转移力度不断增加,可以说,目前,我国纺织业无论是产品设计还是贸易营销,两端正进一步国际化甚至引领国际潮流。因此,产业对从业人员提出"能用英语就专业问题进行基础而有效的沟通"的新要求。高职类纺织专业毕业生既要有一定的纺织技术技能又要有一定的英语素养。另外,我国纺织服装产业的巨大优势,吸引了周边国家的学子来华求学,特别是"一带一路"沿线国家来我国高职院校求学的留学生人数逐年递增。留学生教育一般以英语为媒介,开展专业教育,同时又必须提高其汉语水平,是国内高职院校需要面对的新问题。基于市场需求,编写《纺织导论双语教程》以期能较好地服务国内"走出去"的高职类纺织专业毕业生,又能为来华留学生所用。本书以纺织产品加工产业链为主线,从纺织专业外贸相关工作出发,统筹考虑,选出具有代表性和典型性的常用纺织知识点,凝练出关键的纺织专业英文词汇、英文例句,力求"实用、有效"。本书具有以下特色。

1.以高职学生使用为主,留学生使用为辅,同时注重英语和汉语的简练、规范。高职类纺织专业学生英语基础较薄弱,适用简练、规范的英语,而来华留学生除学习我国纺织服装先进的专业知识外,另一目标是学习汉语,加强语言交流,有必要在专业导论课上为其提供标准、规范的汉语学习资源。

2.内容选择上结合高职学生特色,弱化专业理论深度,突出专业实践及其与英语专业知识的结合,吸收纺织导论新型国际化知识,体现"简单、够用、实用"的特点,采用双语(汉语与英语)一体化教学方式,实现学生双语能力的有机结合。

3.全书以"学习—尝试—思考—文化浸润"为宗旨进行教材编写。其中 Learning 部分以"专业导论文本【Text】—英文关键词【Key Words】—重点英文关键句【Key Sentences】"的模式编写,改革了目前市场上大部分中英文双语教材英汉交替翻译的编写模式,Trying 部分选取典型的专业英语实践题,使学生融入简单的实践,既可激发学生对本专业的学习热情,又能锻炼学生的英语表达;Thinking 部分主要提出一些发散性思维问题,或者将项目导论知识点与日常生活交叉结合,使学生对专业导论的拓展有深刻的思考。

本书由赵磊、陈宏武担任主编,王可、陈贵翠担任副主编,陆晓波、徐帅、周红涛、秦晓、王前文、朱挺、陈燕、赵菊梅、毛雷参与编写。全书由刘华、张圣忠审稿,刘华主审。编写工作得到了江苏省高校品牌专业项目、江苏省产教深度融合实训平台项目、江苏省高校"青蓝工程"优秀教学团队项目，以及盐城工业职业技术学院校级教改课题等项目的资助。

本书在编写过程中参考了许多相关图书和论文资料，在此特向这些文献资料的作者致以真挚的谢意！同时对江苏悦达棉纺有限公司、无锡大耀集团、苏州立新集团等单位的外贸专家、专业技术骨干在教材编写过程中给予的大力支持表示感谢！

由于编者水平有限，书中难免存在错误和不当之处，敬请读者批评指正。

编 者

2019 年 7 月

Contents(目录)

Project One   General Introduction(绪论) ·· 1

Project Two   Textile Fiber(纺织纤维) ·· 17
    Task One   Natural Cellulosic Fibers(天然纤维素纤维) ······························· 17
    Task Two   Natural Protein Fibers(天然蛋白质纤维) ··································· 31
    Task Three   Chemical Fibers(化学纤维) ·· 41

Project Three   Yarns and Spinning Technology(纱线与纺纱技术) ················ 66
    Task One   Yarns(纱线) ·· 66
    Task Two   Opening and Cleaning Technology(开清棉技术) ······················· 73
    Task Three   Carding and Combing Technology(梳棉与精梳技术) ················ 83
    Task Four   Drawing and Roving Technology(并条与粗纱技术) ················· 96
    Task Five   Spinning Technology(细纱技术) ··· 102
    Task Six   Winding Technology(络筒技术) ·· 111

Project Four   Woven Fabrics and Woven Technology(机织物与机织技术) ········ 119
    Task One   Warping Technology(整经技术) ·· 119
    Task Two   Warp Sizing Technology(浆纱技术) ··· 125
    Task Three   Woven Technology(织造技术) ··· 132
    Task Four   Woven Fabrics(机织物) ·· 142

Project Five   Knitted Fabrics and Knitting Technology(针织物与针织技术) ········ 155
    Task One   Knitting Fabrics(针织物) ·· 155
    Task Two   Knitting Technology(针织技术) ·· 160

Project Six   Nonwoven Fabrics and Processing Technology(非织造物与非织造技术) ······ 173
    Task One   Nonwoven Fabrics(非织造物) ·· 173
    Task Two   Netting Technology(纤维成网技术) ··· 180
    Task Three   Reinforcement Technology(纤网加固技术) ······························ 190
    Task Four   Nonwoven Technology of Polymer Extrusion(聚合物挤压技术) ······ 208

Project Seven　Textile Dyeing and Finishing(纺织品染整) ································ 214
　　Task One　Pretreatment Technology(前处理技术) ·································· 214
　　Task Two　Dyeing and Printing Technology(染色与印花技术) ················ 220
　　Task Three　Finishing Technology(整理技术) ·· 231

参考文献 ·· 254

Project One　General Introduction(绪论)

Part 1　*Learning*

【Text】

中国的纺织与印染技术具有非常悠久的历史。早在原始社会时期,古人为了适应气候的变化,已懂得就地取材(纺织材料),如将采集的野生的麻、蚕丝及猎获的鸟兽毛羽等作为纺织和印染原料,采用手工方法制成粗陋的衣服以取代蔽体的草叶和兽皮[1]。直至今天,日常的衣服、一些家居用品和艺术品都是纺织和印染技术的产物[2]。可以说人们的生活离不开纺织[3]。

(注 [1,2,3]对应下文的【Key Sentences】中的序号,全书同。)

【Key Words】

primitive [ˈprimətiv] 原始的,远古的
ancient [ˈeinʃənt] 古代人
hemp [hemp] 大麻
silk [silk] 蚕丝

shield [ʃi:ld] 掩护物,遮挡物
household supplies 家居用品
inseparable [inˈseprəbl] (与某事物)不可分离的,分不开的

【Key Sentences】

1. As early as the **primitive** social period, in order to adapt to climate change, the **ancients** understood how to use local (textile) materials such as wild **hemp**, **silk**, hunted birds' feathers and animal skins as raw materials of textile and printing, to make rough clothes by hand and replace the grass leaves and animal skins as the **shield**.

2. Today, our daily clothes, **household supplies** and artworks are the products of textile and dyeing techniques.

3. It can be said that people's lives are **inseparable** from textiles.

【Text】

纺织品对日常生活有着非常重要的影响,每个人都需要加以了解。纤维是构成纺织品最基本的单元,其中最常见的纤维是棉纤维、麻纤维、羊毛纤维、蚕丝纤维[1];在购买纺织品时,也会发现产品吊牌中会标示涤纶、氨纶、锦纶等合成纤维,或者莫代尔、天丝、牛奶蛋白等新型再生纤维[2];当然也会在各种媒体上听闻碳纤维、玻璃纤维、芳纶等高性能纤维。

在纺织科学中,纺织品被自由定义为任何由纤维制成的产品[3];该术语不仅指机织面料,还指针织面料、非织造布和特殊结构织物。由纤维制成的纺织品和织物用途广泛,并且不断发现新的应用。人们对纤维在服装和家用纺织品中的应用耳熟能详[4]。事实上,纤维在各种活动或环境中都扮演重要的角色,从在沙上玩耍到在月球上行走[5]。

【Key Words】

cotton [ˈkɔtn] 棉,棉花
wool [wul] 羊毛
synthetic [sinˈθetik] 人工的,合成的
polyester [ˌpɔliˈestə(r)] 聚酯纤维,涤纶
spandex [ˈspændeks] 氨纶
nylon 聚酰胺纤维,尼龙
regenerated [riˈdʒenəreitid] fiber 再生纤维
modal 莫代尔纤维
tencel 天丝
milk protein fiber 牛奶蛋白纤维

high-performance fiber 高性能纤维
carbon [ˈkaːbən] fiber 碳纤维
glass fiber 玻璃纤维
aramid [ˈɛrəmid] fiber 芳纶
be defined as 被定义为
woven fabric 机织物
knitted fabric 针织物
nonwoven fabric 非织造织物
household textile 家用纺织品
essential [iˈsenʃl] 必不可少的,完全必要的

【Key Sentences】

1. Fiber is the most basic component of textiles, and the most common fibers are **cotton** fiber, hemp fiber, **wool** fiber, and silk fiber.

2. When purchasing textiles, we also find that the producttag **tags** labeled with **synthetic** fibers such as **polyester**, **spandex** and **nylon**, or new **regenerated fibers** such as **modal**, **tencel** and **milk protein fiber.**

3. In textile science, textile is freely **defined as** any products made from fibers.

4. The uses of fibers in clothing and **household textiles** are familiar to everyone.

5. In fact, fibers play their **essential** roles in nearly all kinds of activities or situations, from playing in the sand to walking on the moon.

【Text】

(纺织品的加工)一般是先将纤维加工成纱线,纱线再经过加工变成织物,最后织物再经加工变成终端服装[1]。纤维作为原材料,要么纺(或加捻)成纱线,要么直接压缩成织物[2];纱线作为纤维的排列,用机织、针织技术或以其他方式制成织物[3];织物作为纤维或纱线的后续产物,通过各种整理工艺成为消费者商品。可见,原材料向最终消费者商品的发展是合乎逻辑的。研究纺织材料从"纤维到纱线到织物"的加工过程将有助于我们熟悉了解织物的结构和最终的品质[4]。

【Key Words】

spun (spin 的过去分词) [spʌn] 纺纱
twist [twist] 加捻
compress [kəmˈpres] 压缩
arrangement [əˈreindʒmənt] 安排,排列

logical [ˈlɔdʒikl] 合乎常理的,符合逻辑的
sequence [ˈsiːkwəns] 顺序,次序
construction [kənˈstrʌkʃn] 构造,结构
familiar [fəˈmiliə(r)] 熟悉的,通晓的

【Key Sentences】

1. Generally, fibers are first processed into yarns, which then are processed into fabrics, and finally, the fabrics are processed into terminal garments.
2. Fibers as raw materials are either **spun** (or **twisted**) into yarn or directly **compressed** into fabric.
3. Yarns as **arrangement** of fibers are made into fabric by woven technology, knitting technology, or other means.
4. Studying the processing of textile materials—"fiber to yarn to fabric" will help us understand the **construction** and ultimate qualities of the fabrics.

【Text】

狭义的纺织是指纺纱和织布;广义的纺织则还包括原料初加工、缫丝、染整、化学纤维制造和纺织贸易等[1]。纺织品可以满足人们基本的衣着需要,但随着社会的进步与物质的极大丰富,人们对纺织品提出了更多、更高的的要求[2];纺织品的应用领域越来越广,由原来的服装用纺织品与装饰用纺织品逐渐拓展到产业用纺织品[3]。

服装用纺织品包括制作服装的各种纺织面料和缝纫线、松紧带、领衬等各种纺织品辅料[4]。装饰用纺织品指室内用品、床上用品和户外用品。室内用品包括家具、餐厅和浴室用品,如地毯、沙发套、坐垫(套)、壁毯、贴布、罩、窗帘、毛巾等[5];床上用品包括床罩、床单、被子、被套、枕芯、被芯、枕套等;户外用品包括人造草坪等。装饰用纺织品在品种结构、织纹图案和配色等方面较其他纺织品有更多的突出特点[6]。

【Key Words】

narrow ['nærəu] 狭义的
generalized ['dʒenrəlaizd] 广义的
reeling silk 缫丝
enrichment [in'ritʃmənt] 丰富,充实
put forward 提出
apparel [ə'pærəl] 衣服,服装
extensive [ik'stensiv] 广泛的
decoration [ˌdekə'reiʃn] 装饰
extend [ik'stend] 扩大,扩展
industrial textile 产业用纺织品
apparel textiles 服装用纺织品
accessories [ək'sɛsəriz] 辅料
sewing threads 缝纫线
elastic band 松紧带
collar lining 领衬

decorative textiles 装饰用纺织品
indoor supplies 室内用品
bedding supplies 床上用品
outdoor supplies 户外用品
bathroom ['ba:θru:m] 浴室,洗手间
sofa cover 沙发套
cushion ['kuʃn] 坐垫,靠垫
tapestries ['tæpistriz] 壁毯
stickers ['stikəz] 标签布
bedspread 床罩
sheet 床单
quilt [kwilt] 被子,棉被
duvet ['du:vei] cover 被套,被罩
pillow core 枕芯
quilt core 被芯

pillowcase 枕套
artificial [ˌɑːtiˈfiʃl] turf [tɜːf] 人工草皮
prominent [ˈprɔminənt] 显著的,突出的

feature [ˈfiːtʃə(r)] 特点
pattern [ˈpætn] 花型,图案

【Key Sentences】

1. **Generalized** textiles include raw material processing, **reeling silk**, dyeing and finishing, chemical fiber manufacturing and textile trade.

2. Textiles can meet people's basic clothing need, but with the advancement of society and the great **enrichment** of materials, people have **put forward** more and higher requirements for textiles.

3. The application fields of textiles are becoming more and more **extensive**, and the original textiles for **apparel** and **decoration** are gradually **extended** to **industrial textiles.**

4. Apparel textiles include various fabrics for making garments and various textile **accessories** such as **sewing threads**, **elastic bands**, and **collar linings**.

Fig 1.1 shows some apparel textiles.

(a) Cotton fabric (棉织物)　(b) Hemp fabric (麻织物)　(c) Wool fabric (毛织物)　(d) Silk fabric (丝织物)
(e) Polyester fabric (涤纶织物)　(f) Nylon fabric (锦纶织物)　(g) Tencel fabric (天丝织物)　(h) Milk protein fiber fabric (牛奶蛋白纤维织物)
(i) Sewing thread (缝纫线)　(j) Elastic band (松紧带)　(k) Lining cloth (衬布)　(l) Clothing lace (服装花边)

Fig 1.1　Some apparel textiles(部分服装用纺织品)

5. **Indoor supplies** include furniture, dining room and **bathroom** items such as carpets, **sofa covers**, **cushions** (sets), **tapestries**, **stickers**, covers, curtains, towels.

Fig 1.2 shows some decorative textiles.

（a）Cushion（坐垫）　　（b）Bath towel（浴巾）　　（c）Wall cloth（墙布）　　（d）Dining chair/table cloth（餐椅布、台布）

（e）Bed cover/pillowcase/curtain（床罩、枕套、窗帘）　（f）Suit dust cover（西服防尘罩）　（g）TV dust cover（电视防尘罩）　（h）Air conditioning dust cover（空调防尘罩）

（i）Non-woven bag（非织造布袋）　（j）Blanket（毛毯）　（k）Lampshade cloth（布艺灯罩）　（l）Artificial grass（人造草皮）

Fig 1.2　Some decorative textiles（部分装饰用纺织品）

6. Decorative textiles have more **prominent features** than other textiles in product structure, **pattern** and color matching.

【Text】

产业用纺织品是经专门设计的且具有特定功能的纺织品。它可应用于医疗卫生、环境保护、土木建筑、交通运输、安全防护、航空航天、新能源、农牧渔业等领域[1]。

医疗与卫生用纺织品,通常指应用于医疗、防护、保健及卫生用途的纺织品。医疗用纺织品通常具有防病毒、防渗透、抗菌、抗静电等功能,可以保护医护人员,减少患者感染的概率[2];另一类医疗用纺织品直接用于医疗操作或植入人体,植入人体的纺织品属于高科技产品[3];目前使用最成熟的医疗用纺织品就是可溶性医用缝合线[4]。卫生用纺织品主要用于家庭清洁和个人卫生护理领域,大多是一次性使用,一般由非织造布经后整理而成[5],生活中常用的有尿不湿、湿纸巾、卫生巾等。

【Key Words】

medical care 医疗
hygiene ['haidʒi:n] 卫生
civil construction 土木建筑
aerospace ['eərəuspeis] 航空航天
agriculture ['ægrikʌltʃə(r)] 农业
animal husbandry ['hʌzbəndri] 畜牧业
fishery ['fiʃəri] 渔业
anti-virus ['ænti 'vairəs] 防病毒
anti-infiltration [ˌinfil'treiʃən] 防渗透
anti-bacterial [bæk'tiəriəl] 抗菌
anti-static 抗静电

waist [weist] belt （护）腰带
infection [in'fekʃn] 传染,感染
soluble ['sɔljəbl] 可溶性的
suture ['su:tʃə(r)] （尤指手术后伤口的）缝合线
disposal 一次性的
diaper ['daipə(r)] 婴儿尿片
wet tissue 湿纸巾
sanitary ['sænətri] napkin 卫生巾
valve [vælv]（心脏或血管的）瓣(膜)
vessel ['vesl]（人或动物的）血管,脉管

【Key Sentences】

1. **Industrial textiles** can be used in **medical care** and **hygiene**, environmental protection, **civil construction**, transportation, safety protection, **aerospace**, new energy, **agriculture**, **animal husbandry and fishery** and other fields.

Fig 1.3 shows the classification of industrial textiles.

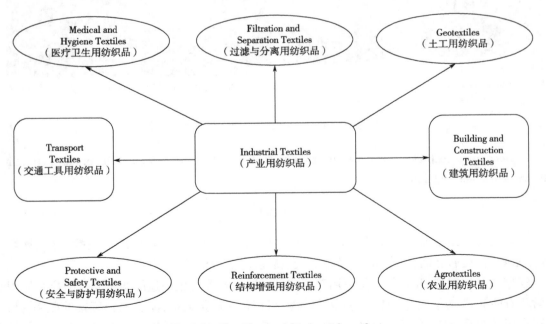

Fig 1.3 Classification of industrial textiles

2. Medical care textiles usually have **anti‑virus**, **anti‑infiltration**, **anti‑bacterial**, **anti‑static** and other functions, which can protect medical workers and reduce the chance of **infection** in patients.

Fig 1.4 shows some medical care and hygiene textiles.

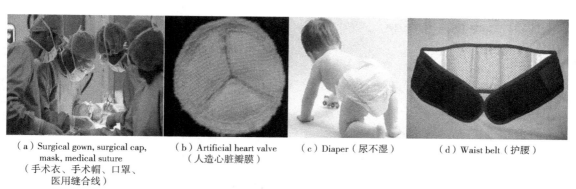

(a) Surgical gown, surgical cap, mask, medical suture（手术衣、手术帽、口罩、医用缝合线）　　(b) Artificial heart valve（人造心脏瓣膜）　　(c) Diaper（尿不湿）　　(d) Waist belt（护腰）

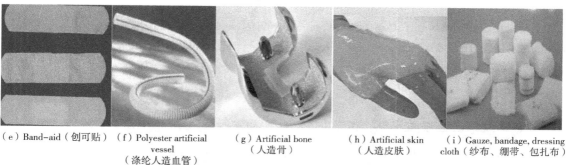

(e) Band-aid（创可贴）　　(f) Polyester artificial vessel（涤纶人造血管）　　(g) Artificial bone（人造骨）　　(h) Artificial skin（人造皮肤）　　(i) Gauze, bandage, dressing cloth（纱布、绷带、包扎布）

Fig 1.4　Some medical care and hygiene textiles（医疗卫生用纺织品）

3. Another type of medical care textiles is directly used for medical operations or **implanted** in the human body.

4. Currently, the most mature medical care textiles is **soluble** medical **sutures**.

5. Hygiene textiles generally made of finished nonwoven fabrics are mainly used in the field of household cleaning and personal hygiene, mostly for **disposal** use.

【Text】

过滤与分离用纺织品通常指应用于气/固分离、液/固分离、气/液分离、固/固分离、液/液分离、气/气分离等领域的纺织品[1]。如食品、医药和化工行业用高纯水预过滤，果汁、饮料、酒类过滤，大型建筑送风系统的空气过滤，汽车的空气、燃油、机油过滤，肾脏患者血液透析时的杂质分离，工业高温烟气过滤等。这些过滤材料都是使用特殊纤维制成的。最熟悉的过滤用纺织品就是口罩，每个雾霾天都离不开它[2]。

土工用纺织品(简称为土工布)，通常指由各种纤维材料通过机织、针织、非织造和复合等加工方法制成[3]；土工布是在岩土工程和土木工程中用于与土壤和(或)其他材料相接触使用的纺织品的总称，可实现隔离、过滤、增强、防渗、防护和排水等功能[4]。

【Key Words】

filtration and separation textiles 过滤与分离用纺织品
beverage ['bevəridʒ] 饮料
alcohol ['ælkəhɔl] 酒, 酒精
fuel ['fju:əl] 燃油
oil [oil] 机油
automobile ['ɔ:təməbi:l] 汽车
impurity [im'pjuərəti] 杂质
hemodialysis [hi:mədai'ælisis] 血液透析, 血液渗析

kidney ['kidni] 肾, 肾脏
filter 过滤器
mask 口罩
hazy ['heizi] 薄雾蒙蒙的
geotechnical textile = geotextile 土工用纺织品
geotechnical engineering 岩土工程
civil engineering 土木工程
seepage prevention 防渗
drainage ['dreinidʒ] 排水, 放水

【Key Sentences】

1. **Filtration and separation textiles** generally refer to textiles used in the fields of gas/solid separation, liquid/solid separation, gas/liquid separation, solid/solid separation, liquid/liquid separation, gas/gas separation, and so on.

Fig 1.5 shows some filtration and separation textiles.

(a) Water filter materials (水过滤材料)

(b) Air filter materials (空气过滤材料)

(c) Anti-haze filter mask (防雾霾过滤口罩)

(d) Fuel, oil filter and high temperature air filter materials (燃油、机油过滤材料及高温空气过滤材料)

Fig 1.5 Some filtration and separation textiles(部分过滤与分离用纺织品)

2. The most familiar **filter** textiles is **mask**, which is indispensable in each **hazy** day.

3. **Geotechnical textiles** (sometimes simply called **geotextiles**) usually refer to textiles made of various fiber materials through weaving, knitting, nonwoven and composite processing methods.

4. Geotextile is a general term for textiles used in contact with soil and/or other materials for **geotechnical and civil engineering**, and it can achieve the functions of isolation, filtration, enhancement, **seepage prevention**, protection and **drainage**.

Fig 1.6 shows some geotextiles.

(a) Geotextile for soil erosion
（护岸、护坡防泥土流失用土工布）

(b) Coated geotextile for isolation
（隔离涂层土工布）

(c) Geotextile for pavement filtration
（路面过滤用土工布）

(d) Soft permeable pipe
（软式透水管）

(e) Geotextile mat（土工网垫）

(f) Unidirectional tensile geogrid
（聚丙烯单向拉伸土工格栅）

Fig 1.6　Some geotextiles(部分土工用纺织品)

【Text】

交通工具用纺织品,通常指在汽车、火车、船舶、飞机等交通工具中应用的纺织品[1]。主要包括轮胎帘子布、内饰织物、安全带和安全气囊、填充用纺织品等[2]。

安全与防护用纺织品,通常指在特定的环境下保护人员和动物免受物理、生物、化学和机械等因素伤害的纺织品[3]。它包括防割、防刺、防弹、防爆、防火、防尘、防生化、防辐射等功能[4]。安全及防护用纺织品又可分为个体防护装备和防护服两大类,这些制品广泛用于石油、冶金、发电、水利、矿山、海运、消防、救生等方面。国家日益重视劳动安全,该类纺织品的研发力度越来越大,今后将成为产业用纺织品的重要分支[5]。

【Key Words】

transport textiles　交通工具用纺织品
aircraft ['eəkrɑːft]　飞机
transport　交通工具,运输工具
tire cord fabric　轮胎用帘子布
seat belt　（汽车或飞机上的）安全带
airbag　安全气囊
safety and protection textiles　安全与防护用纺织品
biological [ˌbaiə'lɔdʒikl]　生物学的,生物的
anti-cutting　防割
stab-resistant　防刺
bulletproof　防弹
explosion-proof　防爆
fireproof　防火

dustproof　防尘
anti-biochemistry　防生化
radiation-proof　防辐射
personal protective equipment　个人防护装备
protective clothingd　防护服
petroleum [pə'trəuliəm]　石油,原油
metallurgy [mə'tælədʒi]　冶金
power generation　发电
water conservancy [kən'səːvənsi]　水利
mining　采矿
shipping　海运
lifesaving　救生
labor safety　劳动安全
branch　分支

【Key Sentences】

1. **Transport textiles** usually refer to textiles used in automobiles, trains, ships, **aircraft** and other **transports**.

Fig1.7 shows some transport textiles.

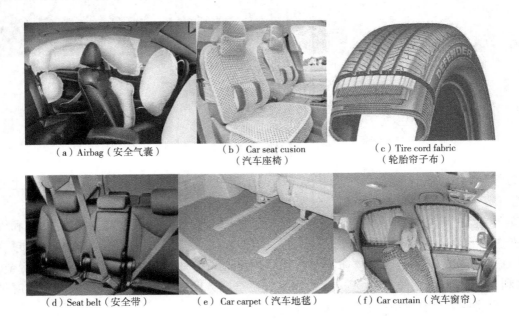

(a) Airbag（安全气囊）　　(b) Car seat cusion（汽车座椅）　　(c) Tire cord fabric（轮胎帘子布）

(d) Seat belt（安全带）　　(e) Car carpet（汽车地毯）　　(f) Car curtain（汽车窗帘）

Fig 1.7　Some transport textiles（交通工具用纺织品）

2. They mainly include **tire cord fabrics, interior fabrics, seat belts and airbags, filling textiles and so on.**

3. **Safety and protection textiles** usually refer to textiles that protect people and animals from physical, **biological**, chemical and mechanical factors in a specific environment.

Fig 1.8 shows some safety and protection textiles.

（a）Anti-biochemical clothing（防生化服）　　（b）Diving suit（潜水服）　　（c）Body armour（防弹衣）

（d）Life vest（救生衣）　　（e）Strab-resistant clothing（防刺服）　　（f）Antiexposure suit（防爆服）

Fig 1.8　Some safety and protection textiles（安全与防护用纺织品）

4. It includes **anti-cutting, stab-resistant, bulletproof, explosion-proof, fireproof, dustproof, anti-biochemistry, radiation-proof** and other functions.

5. The country is paying more and more attention to **labor safety**, so the research and development of such textiles is increasing, and they will become an important **branch** of industrial textiles in the future.

【Text】

结构增强用纺织品指用于（纺织）复合材料中作为增强骨架材料的纺织品，包括短纤维、长丝、纱线以及各种织物和非织造物[1]。现代生活中，飞机、船舱、汽车、火车、文体用品、化工、风力发电等领域都大量采用了纺织品，在结构增强的同时，减轻重量、降低成本、节约能源[2]。撑竿跳高杆、钓鱼竿、棒球杆等制品是比较熟悉的结构增强纺织品。

建筑用纺织品通常指应用于长久性或临时性建筑物和建筑设施的纺织品，主要用于增强、修复、防水、隔热、吸音、隔音、视觉保护、防日晒、耐酸碱腐蚀、减震等，体现了安全、环保节能和舒适等特点[3]。建筑用防水卷材主要用于建筑墙体和屋面,起到抵御外界雨水、地下水渗漏的作用[4];在建筑物的外墙和内墙使用产业用纺织品,可以达到轻质、阻燃、增强、保温、美观等效果[5]。

【Key Words】

structural reinforcement textiles 结构增强纺织品
framework ['freimwɜːk] (物体的)骨架
staple fiber 短纤维
filament ['filəm(ə)nt] 长丝
cabin ['kæbinz] 轮船或飞机的座舱
stationery ['steiʃənri] sports goods 文体用品
chemicals 化工
wind power generation 风力发电
pole vault pole 撑杆跳杆
fishing rod 钓鱼竿
baseball bat 棒球杆
building and construction textiles 建筑用纺织品
permanent ['pəːmənənt] 永久的,长久的
temporary ['temprəri] 短暂的,暂时的
repair [ri'prə] 修复

waterproofing 防水
heat insulation 隔热
sound absorption 吸声
sound insulation 隔声
visual [vi'ʒuəl] 视觉的
anti-sun 防日晒
acid and alkali corrosion [kə'rəuʒn] resistance 耐酸碱腐蚀
shock absorption 减震
membrane ['membrein] (可起防水作用的)膜状物
roof 屋顶
leakage ['liːkidʒ] 泄漏,渗漏
groundwater 地下水
yacht [jɔt] 帆船,游艇,快艇
stadium ['steidiəm] 体育场运动场
ceiling ['siːliŋ] 顶棚
substrate ['sʌbstreit] 底层,基底,基层

【Key Sentences】

1. **Structural reinforcement textiles** are used in composites as reinforcing framework materials, including **staple fibers**, **filaments**, yarns, and various fabrics and nonwovens.

Fig 1.9 shows some structural reinforcement textile composites.

2. In modern life, aircrafts, **cabins**, automobiles, trains, **stationery sports goods**, **chemicals**, **wind power generation** and other fields have adopted a large number of textiles to lose weight, reduce costs and save energy while strengthening the structure.

3. **Building and construction textiles** generally refer to textiles used in **permanent** or **temporary** buildings and building facilities, mainly for reinforcement, **repair**, **waterproofing**, **heat insulation**, **sound absorption**, **sound insulation**, **visual protection**, **anti-sun**, **acid and alkali corrosion resistance**, **shock absorption**, and other functions, showing the characteristics of safety, environmental protection, energy saving and comfort.

Fig 1.10 shows some building and construction textiles.

4. The waterproof **membrane** for building is mainly used for building walls and **roofs** to resist the **leakage** of rainwater and **groundwater**.

Project One General Introduction(绪论)

Fig 1.9 Some structural reinforcement textile composites(结构增强纺织复合材料)

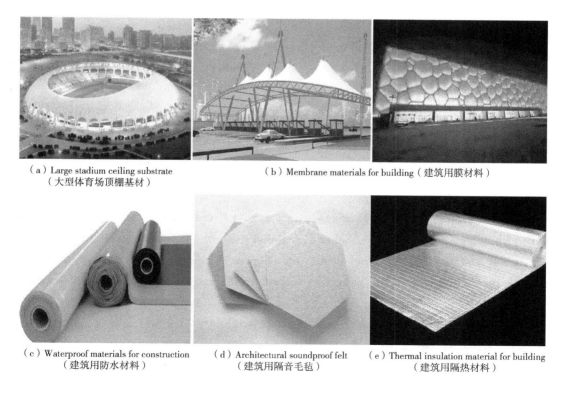

Fig 1.10 Some building and construction textiles(建筑用纺织品)

5. The use of industrial textiles on the outer and inner walls of buildings can achieve **light weight**, **flame retardant** and reinforcement, (thermal) insulation, beauty and other effects.

【Text】

农业用纺织品通常应用于农业耕种、园艺、森林、畜牧、水产养殖及其他农、林、牧、渔业活动[1]。它们有助于提高农产品产量,减少化学药品用量,包含动植物生长、防护和储存过程中使用的所有纺织品。

21世纪,产业用纺织品对于纺织工业具有重大意义,它突破服装用纺织品的垄断格局,引领了纺织革命的创新[2];它不但体现了学科交叉的优势,而且获得了较好的经济和社会效益,具有非常广阔的发展前景[3]。

因此,随着纺织加工技术的进步以及人们对高品质纺织品的个性化需求,要求人们研究并适应高速发展的现代纺织技术[4]。现代纺织技术就是解决现代纺织生产、管理等实践问题的方法和技艺,以推进纺织新技术、智能化改造传统纺织、特种功能纺织品等领域的发展[5]。因此,现代纺织技术涉及纺织工艺、纺织设备、纺织信息化、纺织生产、纺织后加工以及纺织贸易等多个领域。

【Key Words】

agricultural [ˌæɡrɪkʌltʃərəl] textiles 农业用纺织品
cultivation [ˌkʌltɪˈveɪʃn] 耕种,种植
horticulture [ˈhɔːtɪkʌltʃə(r)] 园艺
forestry [ˈfɒrɪstri] 林业
animal husbandry 畜牧业
aquaculture [ˈækwəkʌltʃə(r)] 水产养殖
output 产量
break through 冲破,突破

monopoly [məˈnɒpəli] pattern 垄断格局
embody [ɪmˈbɒdi] 体现
interdisciplinary [ˌɪntədɪsəˈplɪnəri] 多学科的,跨学科的
individualized [ˌɪndɪˈvɪdʒuəlaɪzd] 个性化的
intelligent reformation 智能化改造
informationization 信息化

【Key Sentences】

1. **Agricultural textiles** usually are used in agricultural **cultivation**, **horticulture**, **forestry**, **animal husbandry**, **aquaculture** and other agricultural, forestry, husbandry, fishery activities.

2. In the 21st century, industrial textiles are of great significance to textile industry, because they **break through** the **monopoly pattern** of apparel textiles and lead the innovation of textile revolution.

3. They not only **embody** the advantage of **interdisciplinary**, but also achieve better economic and social benefits, so they have a very broad development prospect.

4. Therefore, with the advancement of textile processing technology and the **individualized** demand for high-quality textiles, it requires people to research and adapt to the rapid development of modern textile technology.

5. Modern textile technology is a method and skill to solve practical problems in modern textile production and management, to promote the development of new textile technologies, **intelligent reformation** of traditional textile technology, special functional textiles, etc.

Fig 1.11 Shows some agriculture textiles.

(a) Thermal insulation membrane(保温膜)　(b) Insect-resistant sack(防虫套袋)　(c) Blackout-fabric(遮光布)

(d) Insectproof net(防虫网)　(e) Colourful strip fabric(彩条布)　(f) Plastic woven bag(蛇皮袋)

Fig 1.11　Some agricultural textiles(农业用纺织品)

【Text】

本书将根据纤维到织物成品的系列加工链介绍纺织导论的相关知识[1]。简单介绍常用纺织纤维的种类与性能特点,纱线的种类及其性能表征[2];传统环锭纺纱加工的基本原理、加工流程及主要设备,新型纺纱的加工原理及其特点;机织物的基本组织及其性能特点,机织加工的基本原理、流程、过程及主要设备;针织物常用组织、加工原理、过程及主要设备;非织造织物加工的基本原理、过程及主要设备;以及纺织品染整加工的原理及其作用,特别是一些新型的功能整理开发与应用。这些纺织导论知识的学习能为后续专业课程的详细学习打下坚实的基础[3]。

【Key Words】

textile introduction　纺织导论
processing chain　加工链
traditional ring spinning　传统环锭纺
basic weave　基本组织

functional finishing　功能整理
solid foundation　坚实的基础
follow-up　后续的
professional [prəˈfeʃənl] 专业的

course   课程

【Key Sentences】

1. This book will introduce the relevant knowledge of **textile introduction** based on a series of **processing chains** from fiber to fabric.

2. It briefly introduces the types and properties of common textile fibers, the types of yarns and their performance characterization.

3. The learning of these textile introduction can lay a **solid foundation** for the detailed study of the **follow-up professional courses**.

Part 2   *Trying*

1. Watch 2 to 3 famous clothing brand advertisements through online medias and go to the mall for market research, what do you know about these clothing brands from your own point of view?

_____
_____
_____
_____

2. According to your own experience, what aspects should you pay attention to in order to build a good textile brand?

_____
_____
_____
_____

Part 3   *Thinking*

1. What's your new understanding of the textile industry after learning this project? Which industries are related to the textile industry?

_____
_____
_____
_____

2. What is the inspiration for your professional study after learning this project? Talk about your career plan according to your textile major.

_____
_____
_____
_____

Project Two　Textile Fiber(纺织纤维)

Task One　Natural Cellulosic Fibers(天然纤维素纤维)

Part 1　*Learning*

【Text】

　　人类利用棉花的历史相当久远,早在公元前2000多年,人类就开始采集野生的棉纤维用来御寒,后来,棉花逐渐被推广种植。目前,棉纤维是世界上使用最多的纤维,它也是最重要的纤维素纤维之一[1]。然而,这些年,棉纤维的世界用量在逐渐减少。棉纤维要能被纺制成纱线,其必须是成熟的。成熟的棉纤维在长度方向上具有天然转曲,使棉纤维具有良好的抱合力,增加了纤维之间的摩擦力,使棉纤维可形成具有较高强力的短纤纱[2]。正常成熟的棉纤维横截面不是圆形,就像带有中腔的豆子[3]。棉纤维结构从外到内依次是表皮、初生层、次生层和中腔[4]。次生胞壁的厚度是棉纤维成熟的一个重要标志。

【Key Words】

plant 种植
BC 公元前
cotton['kɔtn] 棉花,棉纤维
fiber ['faibə] 纤维
natural ['nætʃrəl] 天然的,自然的
cellulosic [selju'ləusik] 纤维素的
ball 　(植)圆荚,铃
spun [spʌn] (spin 的过去式和过去分词)纺纱
mature [mə'tʃuə(r)] 成熟的
lengthwise ['leŋθwaiz] 纵向的,长度方向的
convolution [ˌkɔnvə'luːʃn] 回旋,盘旋
　(此处译为:转曲)
cohesion [kəu'hiːʒn] 结合力,抱合力
cross-section ['krɔs'sekʃən] 横截面
bean [biːn] 豆子
lumen ['luːmen] 中腔
cuticle layer　表皮层,角质层
primary wall　初生胞壁、初生层
secondary wall　次生胞壁、次生层
indicator ['indikeitə(r)] 指示、标志

【Key Sentences】

1. At present, **cotton** is the fiber used most in the word, which is one of the very important natural cellulosic fibers.

Fig 2.1 shows an open cotton ball.

2. The **mature** cotton has natural **convolution**s along the **lengthwise** of the fibers, which lead to a good **cohesion** and increase the friction between fibers, so cotton fibers form a high-strength spun yarn.

3. The cross-section of normal mature cotton fiber is not round, just like a **bean** with a **lumen**. Fig 2.2 shows the cross section and lengthwise morphology of cotton fiber.

4. From the outside to the inside, the cotton fiber structure is the **cuticle layer**, the **primary wall**, the **secondary wall** and the lumen.

Fig 2.3 shows the cotton fiber structure.

Fig 2.1　Open cotton ball(开花的棉铃)

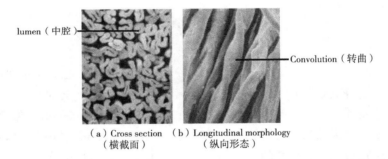

(a) Cross section　(b) Longitudinal morphology
　（横截面）　　　（纵向形态）

Fig 2.2　Cross section and longitudinal morphology of cotton fibe(棉纤维的横截面与纵向形态)

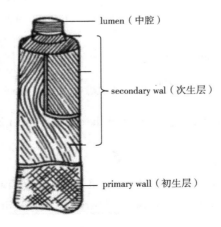

Fig 2.3　Cotton fiber structure(棉纤维的结构)

【Text】

棉纤维的主要组成物质是纤维素。正常成熟的棉纤维的纤维素含量约为94%。此外,还有少量的果胶、蜡质等非纤维素物质。棉纤维耐碱但不耐酸。酸使纤维素分解,大分子链断裂,棉纤维受损导致强度下降[1]。

相对较好的吸湿性和良好的芯吸性使棉纤维成为最舒适的纤维之一,棉纤维的公定回潮率可以达到8.5%[2]。因为纤维素的羟基基团,使得棉花有很强的吸水作用[3]。因此,特别是在夏天,身体的汗液会被棉织物吸收,沿着纱线输送到衣物外表面并蒸发到空气中。因此,这对维持体温有帮助。遗憾的是,棉纤维的亲水性使其容易出现水渍,水溶性色素会随着水分渗入纤维;当水分蒸发后,着色剂会停留在纤维上[4]。也许,棉织物的主要缺点是它们有折皱的倾向,并且难以去除,这是氢键的断裂与重建而使织物保持褶皱,所以,棉织物需要熨烫[5]。

【Key Words】

constituent [kənˈstitjuənt] 成分,组成物质
cellulose [ˈseljuləus] 纤维素
non-cellulosic 非纤维素的
pectin [ˈpektin] 果胶
wax 蜡质
alkali-resistant 耐碱的
acid-resistant 耐酸的
macromolecular chain 大分子链
moisture [ˈmɔistʃ(r)] absorption 吸湿性
wicking [ˈwikiŋ] property 芯吸性(毛细管特性)

official moisture regain rate 公定回潮率
hydroxyl [haiˈdrɔksil] group 羟基
perspiration [ˌpəːspəˈreiʃn] 汗,汗水
susceptible [səˈseptəbl] 易受影响的
water-borne stain [stein] 水渍(污点、色斑、着色剂)
water-soluble colorant [ˈkʌlərənt] 水溶性色素
be trapped in 被捆在……里
wrinkle [ˈriŋkl] 皱褶
rupture [ˈrʌptʃ(r)] 断裂,破裂
iron [ˈaiən] 熨烫

【Key Sentences】

1. Acid make cellulose decompose and **macromolecular chain** break, as a result, cotton fiber damage would result in strength decreasing.

2. The good **moisture absorption and wicking properties** help make cotton one of the more comfortable fibers. Cotton fiber's official moisture regain rate can reach 8.5%.

3. Because of the **hydroxyl groups** in the cellulose, cotton has a high attraction for water.

4. Unfortunately, the **hydrophilic** nature of cotton makes it **susceptible** to **water-borne stains**, **and water-soluble colorants** will penetrate the fiber along with the water; when the water evaporates, the colorant is trapped in the fiber.

5. Perhaps, the major disadvantage to cotton goods is their tendency to **wrinkle** and the difficulty of removing wrinkles. It is the **rupture** and reformation of the hydrogen bonds that helps to maintain

wrinkles, so that cotton goods must be ironed.

【Text】

棉纤维耐光、耐热,棉花自然挂干或烘干机烘干可长时间保持其白度。棉纱线在潮湿环境下比在干燥环境下强度更大,可以解释为:吸收的水分充当润滑剂并赋予纤维更好的弹性[1]。水被吸收时,纱线中的溶胀纤维之间的互相挤压更强烈,内部的摩擦力增强了纱线强度。当水被棉纤维吸收,纤维膨胀,其截面变圆,长度变短,棉织物洗后易收缩[2]。缩水可以通过防缩后整理来控制。耐用性能可以通过化学处理或在棉纤维混纺中使用抗皱纤维(如涤纶)来实现。涤/棉混纺织物为现代顾客提供了无须熨烫的床单和枕套[3]。棉也许比其他任何纤维更能满足服装、家居装饰、休闲等方面的要求[4]。棉具有舒适、在光照下快干、色彩持久、容易打理等优点[5]。

【Key Words】

internal [in'tə:nl] 内部的
lubricant ['lu:brikənt] 润滑剂,润滑油
impart [im'pa:t] 给
flexibility [ˌfleksə'biləti] 弹性,柔韧性
swell [swel] 膨胀,肿胀
susceptible [sə'septəbl] 易受影响的
shrinkage ['ʃriŋkidʒ] 收缩,皱缩,缩水
polyester [ˌpɔli'estə(r)] 聚酯(涤纶)

polyester/cotton blends 涤/棉混纺织物
requirement [ri'kwaiəmənt] 要求,必要条件
apparel [ə'pærəl] (商店出售的)衣服
home furnishing 家居装饰,家居
recreational [ˌrekri'eiʃənl] 娱乐的,消遣的
care for 照顾(此处译:护理)

【Key Sentences】

1. We must know the fact that cotton yarn is stronger when wet than when dry, this is because that the absorbed water acts as an **internal lubricant** which **imparts** a higher level of **flexibility** to the fibers.

2. As water is absorbed, the fiber **swells** and its cross section becomes more rounded and the length becomes more short, so the cotton fabrics are **susceptible** to **shrinkage** after washing. Shrinkage can be controlled by the application of shrink-resistant finishes.

3. **Polyester/cotton blends** provide the modern consumer with no-iron sheets and pillowcases.

4. Perhaps, cotton satisfies the more **requirements** of **apparel**, **home furnishings**, **recreational**, and other uses than any other fiber.

5. Cotton provides many advantages, such as comfort, quick-drying, long-lasting colors, and easy **to care for.**

【Text】

目前,科学家采用现代生物工程技术培育出了彩色棉。彩色棉是指天然生长的非白色棉花,包括棕、绿、红、黄、蓝、紫、灰等多个色泽品种[1]。彩色棉制品有利于人体健康,在纺织过程中减少印染工序,迎合了人类提出的"绿色革命"口号,减少了对环境的污染[2]。彩棉具有舒适、抗

静电、吸汗透气性好、绿色环保等优点。

　　有机棉是在农业生产中,以有机肥(不使用化学肥料)、生物防治病虫害、自然耕作管理为主,从种子到农产品全天然无污染生产的棉花[3]。因此,有机棉花生产是可持续性农业的一个重要组成部分;它对保护生态环境、促进人类健康发展以及满足人们对绿色环保生态服装的消费需求具有重要意义。

　　木棉纤维是一种果实纤维,纤维长 8~32mm,直径为 15~45μm,表面光滑、无天然转曲,截面有大中腔,类似圆形的管状物。中腔的中空率达 80%~90%,远超再生纤维(25%~40%)和其他任何天然材料[4];因此,木棉纤维是超保暖、天然抗菌、不蛀不霉的纺织良材,且超短、超细、超软。

【Key Words】

bioengineering [ˌbaɪəʊˌendʒɪˈnɪərɪŋ] technology　生物工程技术
cultivate　培育
colored cotton　彩棉
brown　棕色
purple [ˈpɜːpl]　紫色的
be conducive to　对……有利的,有帮助的　cater to　迎合
slogan [ˈsləʊɡən]　标语,口号
sweat-absorbent breathable　吸汗透气性
environmental protection　环境保护,绿色环保
organic cotton　有机棉
pollution-free　无污染的,没有污染的
organic fertilizer [ˈfɜːtəlaɪzə(r)]　有机肥

sustainable [səˈsteɪnəbl] agriculture　可持续农业
ecological environment　生态环境
greeneco-friendly [ˌiːkəʊ ˈfrendli]　绿色环保的
kapok fiber　木棉纤维
round tube　圆形管状物
ultra-warm　超保暖的
hollow rate　中空率
mildew-free [ˈmɪldjuː]　不发霉的
antibacterial [ˌæntɪbækˈtɪərɪəl]　抗菌的
insect-free　不被虫蛀的
ultra-short　超短的
ultra-fine　超细的

【Key Sentences】

1. **Colored cotton** refers to non-white cotton that grows naturally, including **brown**, green, red, yellow, blue, **purple**, gray and other color.

Fig 2.4 shows samples of colored cotton.

Fig 2.4　Samples of colored cotton(彩棉样品)

2. Colored cotton products are **conducive** to human health, and reduce printing and dyeing processes in the textile process, so they **cater to** the "green revolution" **slogan** proposed by mankind and reduce environmental pollution.

3. **Organic cotton** is a kind of cotton which is produced naturally and **pollution-free** from seeds to agricultural products, mainly based on **organic fertilizer**(no use of chemical fertilizer), biological control of pests and diseases, and natural farming management in agricultural production.

Fig 2.5 shows kapok fiber fruit and the cross-section morphology of kapok fiber.

 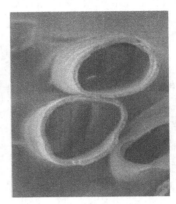

(a) Kapok fiberfruit（木棉果实）　　(b) Gross section morphology of kapok fiber（纤维横截面形态）

Fig 2.5　Kapok fiber fruit and the cross-section morphology of kapok fiber(木棉果实及纤维横截面形态)

4. The **hollow rate** of the lumen is 80%~90%, which far exceeds that of rayon(25%~40%) or any other natural materials.

【Text】

麻纤维是另一种常见的天然纤维素纤维,它是从各类麻植物取得的纤维的统称[1]。麻纤维分茎纤维和叶纤维两类。茎纤维是从麻类植物茎部取得的纤维,因其存在于茎的韧皮部中,所以又称韧皮纤维,绝大多数麻纤维属此类,如苎麻、亚麻、黄麻、大麻等;叶纤维是从麻类植物叶子或叶鞘中取得的纤维,如剑麻、蕉麻等[2]。

亚麻纤维是最古老、最珍贵的天然纤维之一。亚麻纤维素含量高达70%~80%,因此,吸湿性较好。在显微镜下观察,亚麻纤维纵向形态图没有类似棉纤维的天然转曲,但可以看到纵向条纹以及由膨胀和不规则的关节形成的纤维宽度变化的点(称为节点)[3]。这些节点类似于竹子上的关节。其横截面显示了明显的中腔,并呈多边形形状(五角边或六角边)[4]。未成熟的亚麻也许是椭圆形的,并且通常有比成熟纤维更大的内腔。亚麻纤维的干强比棉纤维好,湿强比干强高25%,它因吸收水分而具有较好的舒适性和抗静电性[5]。亚麻比棉纤维长,表面较光滑,光泽好,但亚麻纤维弹性柔性较差,所以,在常规弯曲使用中,亚麻纤维的应用受到限制[6]。亚麻纤维单纤维长度较短,为17~25mm,因此,亚麻纤维需采用工艺纤维(束纤维)纺纱,工艺纤维长度达300~900mm。

【Key Words】

fibrilia 麻纤维
stem fiber 茎纤维
leaf fiber 叶纤维
phloem ['fləuem] 韧皮部
bast fiber 韧皮纤维
ramie ['ræmi] 苎麻,(纺织等用的)苎麻纤维
flax 亚麻
jute 黄麻
hemp 大麻
leaf sheath [ʃi:θ] 叶鞘
sisal ['saisl] 剑麻
abaca 蕉麻
precious ['preʃəs] 珍贵的

microscope ['maikrəskəup] 显微镜
longitudinal [lɔndʒi'tju:dinl]
view [vju:] 纵视图
striation [strai'eiʃn] 条痕,条纹状
joint formation [医]关节形成
node [nəud] 节点
polygonal [pə'ligənl] shape 多边形
static resistance 防静电性,抗静电性
lustrous ['lʌstrəs] 光亮的,有光泽的
pliability [ˌplaiə'biliti] 柔韧性,可弯性
applicability [ˌæplik'biləti] 适用性,适应性
technical fiber 工艺纤维
bundle fiber 束纤维

【Key Sentences】

1. **Fibrilia** is another common natural cellulose fiber, which is a general term for fiber obtained from various types of fibrilia plants.

2. The leaf fiber is the fiber obtained from the leaf or **leaf sheath** of the fibrilia plant, such as **sisal**, **abaca**.

3. The **longitudinal** view of the fiber under the **microscope** shows that there are no convolutions as in cotton, but longitudinal **striations** and the points at which the fiber width changes marked by swelling and irregular **joint formations** called **nodes** can be seen.

Fig 2.6 shows the flax stalks with seeds and Fig 2.7 shows flax technical fiber.

Fig 2.6　Flax stalks with seeds(带种子的亚麻秸秆)

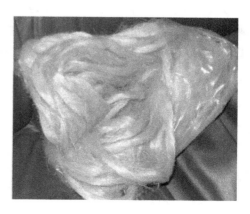

Fig 2.7　Flax technical fiber[亚麻(工艺)纤维]

4. The cross-sectional view clearly shows the lumen, and a somewhat **polygonal shape**. Fig 2.8 shows the cross section and lengthwise of flax.

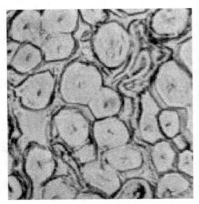

(a) Cross section（横截面） (b) Longitudinal morphology（纵向形态）

Fig 2.8　Cross section and lengthwise of flax（亚麻纤维的横截面与纵向形态）

5. Flax has better dry strength than cotton, and it gets 25 percent stronger that wet strength. It absorbs more moisture for comfort and **static resistance**.

6. It is longer, smoother and more **lustrous** than cotton, but flax has low **pliability** of **flexibility**, so the applicability of the fax, in use of regular bending, is limited.

【Text】

苎麻,通常被称为"中国草",是一种使用较为广泛的韧皮纤维,是早期使用的纺织纤维之一[1]。苎麻在强度、美观和吸湿性方面与亚麻相媲美,但它比亚麻更硬[2]。苎麻单纤维很长很细,长度达120～250 mm,线密度为0.4～0.5tex。苎麻纤维横截面大都呈腰圆形,内有中腔,胞壁有裂纹,纵向大都平直,有横节和竖纹。苎麻外观呈白色,有光泽,几乎呈丝绸状[3]。苎麻的强度优异,但弹性回复率低,伸长率差[4]。苎麻纤维素含量为65%～75%,且苎麻纤维含有多种天然抑菌微量元素;苎麻纤维吸湿透气性是棉纤维的3～5倍[5];苎麻纤维还具有抑菌、透气、凉爽、防腐、防霉、吸汗等功能,是世界公认的"天然纤维之王",最适合做枕套、凉席的面料。苎麻织物也常用于夏季面料和西装面料,是抽纱、刺绣工艺品的优良用布[6]。

【Key Words】

China grass　中国草,即苎麻
rival ['raivl]　与……竞争
silklike　似丝的,丝状的
linear density　线密度
crack　裂纹
cell wall　细胞壁
horizontal node　横节

verticalstriation　竖纹
appearance[ə'piərəns]　外貌,外观
excellent ['eksələnt]　优秀的,卓越的,杰出的
elongation [ˌiːlɔŋ'geiʃn]　延长,伸长
trace element　微量元素
moisture absorption　吸湿

air permeability 透气
anti-bacterial 抑菌
anti-corrosion 防腐
anti-mold 防霉

summer sleeping mat 凉席
drawing and embroidery [imˈbrɔidəri]
craft 抽纱、刺绣工艺品

【Key Sentences】

1. Ramie, normally known as "**China grass**", is a kind of widely used bast fiber and is one of the textile fibers used from early time.

Fig 2.9 shows the phloem of ramie.

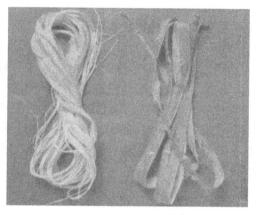

Fig 2.9 Phloem of ramie(苎麻韧皮)

2. Ramie is a minor fiber but can **rival** flax in strength, beauty and absorbency, but it is stiffer than flax.

3. Ramie is white and lustrous and almost **silklike** in **appearance**.

4. The strength of ramie is **excellent**, but elastic recoery is low, and **elongation** is poor.

Fig 2.10 shows the cross section and lengthwise morphology of ramie fiber.

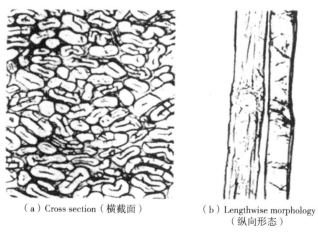

（a）Cross section（横截面）   （b）Lengthwise morphology（纵向形态）

Fig 2.10 Cross section and lengthwise morphology of ramie fiber(苎麻纤维的微观视图)

5. The moisture absorption and air permeability performance of ramie fiber are about 3~5 times that of cotton fiber.

6. Ramie fabrics are also commonly used in summer fabrics and suiting fabrics, and are excellent fabrics for drawn and embroidered crafts.

【Text】

黄麻的纤维素含量高达64%~67%,是典型的纤维素纤维。黄麻纤维吸湿速度快,但吸湿后表面仍保持干燥。黄麻纤维吸湿后膨胀较大并伴随放热现象[1]。黄麻纤维横截面大多呈五角形或六角形,且有呈圆形或卵圆形空腔,有宽有窄[2];黄麻纤维纵向呈竹节状或呈X形节状[3]。黄麻单根纤维非常短,约1~4mm,宽度为10~20μm,必须采用工艺纤维纺纱。但黄麻纤维具有高强度和高初始模量,常用作麻袋、麻布等包装材料及地毯底布等。

随着生态环境问题和能源问题的加剧,开发环保的天然纤维及其相应的纺织材料已成为当前研究的一项热点[4]。如目前新研究与开发的竹原纤维、桑皮纤维、锦葵茎皮纤维、柳皮纤维、木芙蓉韧皮纤维及山麻杆韧皮纤维等。竹原纤维是一种纯天然竹纤维,它是继纤维之后又一具有发展前景的生态功能性纤维,它是从竹材中提取出来的纯粹的天然绿色环保型纤维;竹原纤维纵向有横节,纤维表面有无数微细凹槽。横截面为不规则的椭圆形或腰圆形,有中腔,而且竹纤维横截面上布满了大大小小的空隙,边缘有裂纹[5]。天然竹原纤维具有吸湿、透气、抗菌、抑菌、除臭、防紫外线等良好性能。

【Key Words】

hygroscopic [haigrə'skopik] 吸湿的
expansion [ik'spænʃn] 膨胀
exothermic [ˌeksəu'θəːmik] 放热的
pentagonal [pen'tægənl] 五边形的,五角形的
initial modulus 初始模量
sack 麻袋
burlap ['bəːlæp] 麻布
carpet backing 地毯底布(背衬)
hexagonal [heks'ægənl] 六角形的,六边形的
oval 椭圆形,卵圆形
packaging material 包装材料
bamboo knot 竹子节状
X-shaped knot X形节状
intensification [inˌtensifi'keiʃən] 加剧,增强

bamboo fiber 竹原纤维
mulberry ['mʌbəri] fiber 桑皮纤维
virginica stem bark fiber 锦葵茎皮纤维
willow ['wiləu] fiber 柳皮纤维
hibiscus bast fiber 木芙蓉韧皮纤维
alchornea davidii bast fiber 山麻杆韧皮纤维
numerous ['njuːmərəs] 众多的,许多的
groove [gruːv] 沟槽
ellipse [i'lips] 椭圆形
kidney ellipsoid 腰圆形
be covered with 布满了
gap 空隙
crack 裂纹
deodorization 除臭
anti-ultraviolet 防紫外线

【Key Sentences】

1. Jute fiber has a large swell and is accompanied by an **exothermic** phenomenon after moisture absorption.

2. The cross section of jute fiber is mostly **pentagonal** or **hexagonal**, and has a circular or **oval** lumen with a wide and narrow shape.

Fig 2.11 shows the cross section and lengthwise morphology of jute fiber.

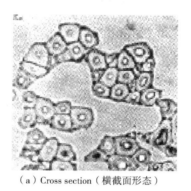
（a）Cross section（横截面形态）

（b）Lengthwise morphology
（纵向形态）

Fig 2.11　Cross section and lengthwise morphology of jute fiber(黄麻纤维的横截面和纵向形态)

3. The lengthwisemorphology of jute fiber has a **bamboo knot** shape or **X-shaped knot**.

4. With the **intensification** of ecological and environmental problems, the development of environmentally friendly natural fibers and their corresponding textile materials has become a hot topic in current research.

5. The cross section of bamboo fiber is covered with large and small **gaps** and **cracks** at the edges.

Fig 2.12 shows the cross section and lengthwise morphology of bamboo fiber.

（a）Cross section（横截面）

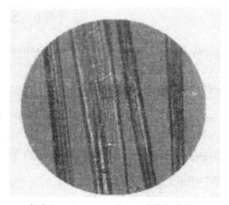
（b）Lengthwise morphology（纵向形态）

Fig 2.12　Cross section and lengthwise morphology of bamboo fiber(竹原纤维的横截面和纵向形态)

【Text】

桑皮纤维是一种新型的野生天然植物纤维,它是从桑皮中提取的纤维素纤维[1]。桑皮纤维既有棉的特征,又具有麻纤维优点,具有极广阔的应用前景。桑皮纤维和人体肌肤有较好的亲和性,吸湿透气性能较好,容易染色,保暖性和光泽良好、手感柔软;桑皮纤维还有较好的抗菌、抑菌性,是高附加值的、可自然降解的环保纤维[2]。但桑皮纤维的纤维素含量较低,仅为40%左右,纤维较短,且强度不够。因此,桑皮纤维常与棉纤维、黏胶纤维等可纺性较好的纤维混纺[3]。

锦葵茎皮纤维具有抗肿瘤、消炎功效,且具有优异的吸湿透气性、良好的抑菌性能,并有可生物降解性[4];它的颜色微黄,手感类似于麻纤维;锦葵纤维素含量约为70%、单纤维长度长,但纤维较硬,纺纱之前需要经过一定的柔软处理。目前,锦葵茎皮纤维主要用于开发纤维增强复合材料[5]。

柳皮纤维是从柳皮中提取出来的一种新型天然纤维素纤维[6],它与麻纤维、竹原纤维有许多相似之处,柳皮纤维与麻类韧皮纤维的化学成分相同。柳树在生长过程中不受污染,因此,柳皮纤维是纯天然原料,且具有绿色环保、舒适透气、瞬间吸水以及抗菌、抑菌等优良特性[7]。这说明柳皮纤维织物的服用性能极佳,保健功效显著。

【Key Words】

affinity [əˈfinəti] 亲和性
high value-added 高附加值的
degradable 可降解的
environmentally friendly 环保的
spinnability 可纺性
anti-tumor [ˈænti'tjuːmə(r)] 抗肿瘤
anti-inflammatory [inˈflæmətri] 消炎的,抗炎的
biodegradability 生物降解性
yellowish [ˈjeləuiʃ] 微黄色的
fiber reinforced composite 纤维增强复合材料
instant water absorption 瞬间吸水
environmental protection 环保
wearability 服用性能
remarkable [riˈmaːkəbl] 显著的,突出的
health care 保健

【Key Sentences】

1. Mulberry fiber is a new type of wild natural plant fiber and it is a cellulose fiber extracted from mulberry.

Fig 2.13 shows the process of mulberry fiber.

2. Mulberry fiber also has good antibacterial properties and it is high value-added, naturally **degradable environmentally friendly** fiber.

3. Therefore, the mulberry fiber is often blended with cotton fiber, viscose fiber and other fibers with good **spinnability**.

4. Virginica stem bark fiber has the effects of **anti-tumor** and **anti-inflammatory**, and has excellent moisture absorption and air permeability, good antibacterial, and **biodegradability**.

5. At present, the virginica stem bark fiber is mainly used for the development of fiber reinforced

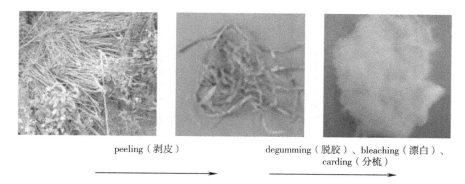

Fig 2.13 Process of mulberry fiber(桑皮纤维的加工)

composites.

6. Willow fiber extracted from the willow skin is a new type of natural cellulose fiber.

7. The willow tree is not polluted during the growth process, therefore, the willow fiber is a pure natural raw material and it has the excellent characteristics of green, comfortable, breathable, **instant water absorption** and antibacterial.

Part 2 *Trying*

1. Quickly distinguish cotton fiber and hemp fiber by visual inspection and touch method.

| Method | cotton | hemp fiber |

2. Observing the difference between cotton and hemp fiber in the longitudinal and transverse directions by electron fiber mirror.

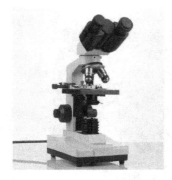

Method	cotton	hemp fiber

3. Describe the burning characteristics of cotton and hemp (including odour, ash, etc.) by combustion method.

Method	cotton	hemp fiber

Part 3 *Thinking*

1. What cotton and linen fabrics have you seen in your daily life? What impressed you with their wearing performance?

_____
_____
_____
_____

2. Go to clothing and home textile stores near the school, and find cotton and linen products.

_____
_____
_____
_____

Task Two　Natural Protein Fibers(天然蛋白质纤维)

Part 1 *Learning*

【Text】

日常生活中,除了经常见到棉、麻等天然纤维素制品外,毛制品也比较常见,如羊毛衫、羊毛西服、羊毛围巾、呢大衣等,这些毛制品是利用羊毛纤维加工而成的。羊毛纤维是一种来源于动物的天然蛋白质纤维[1]。羊毛的生产涉及世界各地许多品种的绵羊,需要大量的牧场。羊毛纤维的横截面是近圆形的,但大部分羊毛纤维的形状是近椭圆形或椭圆形的,羊毛横截面显示出了三个不同部分[2]。外面那层,称为表皮层,由鳞片形成。这些鳞片有点棱角,不规则又互相重叠,它们类似鱼鳞。羊毛纤维的主要部分是皮质层,它延伸至角质层的中心,皮质大概占了纤维重量的90%[3]。羊毛纤维的中心是髓质层,髓质层的尺寸不一,细羊毛可能没有[4]。

【Key Words】

wool [wul] 羊毛
wool sweater　羊毛衫
wool suit　羊毛西服
scarf [skɑːf] 围巾
duffle-coat　呢大衣
protein ['prəutiːn] fiber　蛋白质纤维
breed [briːd] 属,种类
grazing land　类型,牧场
circular ['səːkjələ(r)] 圆形的

slightly elliptical [i'liptikl] 近椭圆形
cuticle ['kjuːtikl] 表皮
scale [skeil] 鱼鳞
horny ['hɔːni] 角状的,有棱角的
overlap [ˌəuvə'læp] 重叠
cortical ['kɔːtikl] 皮质的
cortex ['kɔːteks] 皮质
medulla [mi'dʌlə] 髓质层
invisible [in'vizəbl] 看不见的

【Key Sentences】

1. **Wool** fiber is a natural **protein fiber** and it is from animal.

Fig 2.14 shows the sheep, such as goat, mohair, long-haired rabbit, and yak.

2. The wool fiber section may be nearly **circular**, but most wool fibers' tend to be **slightly elliptical** or oval in shape.

Fig 2.15 Shows the cross section and lengthwise of wool fiber. The cross section of wool can be separated into three parts, as shown in Fig 2.16.

3. The **cortical** cells compose the **cortex** that is the major portion of the wool fiber, cortical cells provide fiber strength and elasticity. The cortex accounts for approximately 90 percent of the fiber mass.

4. The **medulla** is in the center of the fiber, and the size of the medulla varies and in fine fibers may be **invisible**.

(a) Goat（山羊）　　(b) Angora goat（安哥拉山羊）

(c) Long-haired rabbit（长毛兔）　　(d) Yak（牦牛）

Fig 2.14　Four kinds of sheep（四种绵羊）

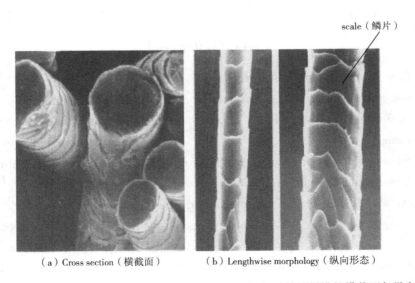

(a) Cross section（横截面）　　(b) Lengthwise morphology（纵向形态）

Fig 2.15　Cross section and lengthwise morphology of wool fiber（羊毛纤维的横截面与纵向形态）

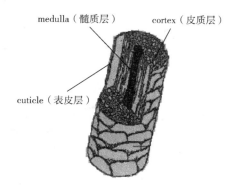

Fig 2.16　Three parts of wool fiber in cross section(羊毛纤维横截面的三个部分)

【Text】

羊毛纤维长度在 3.8~38cm 不等,羊毛的宽度也有很大的差别[1]。细羊毛如美利奴羊毛,平均直径为 15~17μm;而中羊毛平均为 24~34μm,粗羊毛约为 40μm[2]。有些羊毛纤维非常僵硬和粗糙,这些就是所谓的死毛,平均直径约为 70μm。羊毛纤维天然卷曲,有内置的波纹,卷曲增加了纤维的弹性和延伸性,也有助于纱线纺制[3]。羊毛纤维有光泽,细羊毛往往比较粗的羊毛纤维有光泽,天然羊毛纤维的颜色取决于羊的品种[4]。大多数羊毛洗毛后,是淡黄色偏白色或象牙色。有些纤维可能是灰色、黑色、褐色或棕色的。

【Key Words】

fine　纤细的
merino [məˈriːnəu] 美利奴(细毛)羊,美利奴呢绒
micron [ˈmaikrɔn] 微米
medium [ˈmiːdiəm] 中等的,中级的
coarse wool　粗毛
fallen wool　死羊毛,落毛
crimp [krimp] 卷曲
built-in　内置的,固有的

waviness [ˈweivinis] 起浪,成波浪形
luster [ˈlʌstə] 光泽
scouring　冲刷,洗涤
yellowish-white 黄白色
ivory color　乳白色
gray　灰色
tan　黄褐色,棕黄色
brown　棕色的,褐色的

【Key Sentences】

1. Wool fibers vary in length from 3.8 to 38 cm. The width of wool also varies considerably.

2. **Fine** fibers such as **Merino** have an average width of about 15 to 17 **microns**; whereas **medium** wool averages 24 to 34 microns and **coarse wool** about 40 microns.

3. Wool fibers have a natural **crimp**, a **built-in waviness, as shown** in fig 2.17. The crimp increases the elasticity and elongation properties of the fiber and also aids in yarn manufacturing.

4. There is some **luster** to wool fibers. Fine and medium wool tends to have more luster than very

coarse fibers. The color of the natural wool fiber depends on the breed of sheep.

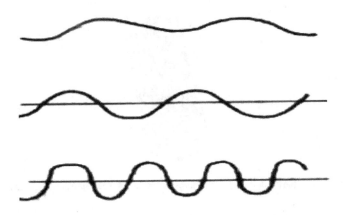

Fig 2.17　Waviness of wool fiber(羊毛纤维的波浪形)

【Text】

干燥时,羊毛的强度可达 0.088~0.150N/tex(1~1.7gf/旦);湿润时,强度下降到 0、0.616~0.132N/tex(7~1.5gf/旦)[1]。羊毛纤维具有优良的弹性和灵活性,在标准条件下,羊毛纤维能延伸 20%~40%[2]。潮湿环境下,它可以延伸到 70%以上,恢复率超强。羊毛的回弹性非常好,它可以在碾压折皱之后恢复原形,然而,通过加热、加湿和施力,持久的折痕或皱纹会在羊毛织物上形成[3]。除了抗压抗皱,羊毛纤维优良的回弹性赋予织物活力,这样形成的通气多孔织物有良好的遮盖力[4]。

羊毛的标准回潮率为 13.6%~16%,饱和条件下,羊毛能吸收其重量的 29%以上的水分[5]。羊毛纤维没有稳定的尺寸。羊毛纤维的结构使其在加工、使用、保养过程中会产生缩绒和毡缩反应[6]。在某种程度上,这是因为纤维的鳞片结构。特别是当毛纤维受到加热、加湿和搅拌处理时,缩绒和毡缩更容易发生[7]。羊毛既可以制成风格各异的四季服装用织物,也可以制成具有特殊要求的工业呢绒、毛毡、衬垫材料,还可以制成壁毯、地毯等装饰品[8]。

【Key Words】

tenacity [təˈnæsəti] 韧性,韧度
grams per denier　克力/旦
extensibility [iksˌtensəˈbiliti] 延伸性
resiliency [riˈziliənsi] 回弹性
crushing　压缩
creasing [kˈriːsiŋ] (使…)起折痕,弄皱
pleat [pliːt] (衣服上的)褶
loft　活力
porous [ˈpɔːrəs] 有毛孔或气孔的
standard moisture regain　标准回潮率

saturation [ˌsætʃəˈreiʃn] 饱和状态,浸透,浸润
dimensionally　在尺寸上
agitation [ˌædʒiˈteiʃn] 搅拌
shrinking　缩绒
felting　毡缩
industrial woolen cloth　工业呢绒
felt　毛毡
cushioning material　衬垫材料
tapestry [ˈtæpəstri] 壁毯

decoration [ˌdekəˈreiʃn] 装饰品

【Key Sentences】

1. The **tenacity** of wool is 1.0 to 1.7 **grams per denier** when dry; when wet, it drops to 0.7 to 1.5 grams per denier.

2. Wool has excellent elasticity and **extensibility**. Under standard conditions, the fiber will extend 20% to 40%.

3. The **resiliency** of wool is exceptionally good. It will readily spring back into shape after **crushing** or **creasing**. However, through the application of heat, moisture and pressure, durable creases or **pleats** can be put into wool fabrics.

4. Besides resistance to crushing and wrinkling, the excellent resilience of wool fiber gives the fabric its **loft**, which produces open, **porous** fabrics with good covering power.

5. The **standard moisture regain** of wool is 13.6 to 16.0 percent. Under **saturation** conditions, wool will absorb more than 29 percent of its weight in moisture.

6. The structure of the fiber contributes to a **shrinking** and **felting** reaction during processing, use and care.

7. When subjected to heat, moisture, and **agitation**, the **shrinking** and **felting** are more likely to occur.

8. Wool can be made into different styles of clothing fabrics for all seasons, or **industrial woolen cloth**, **felt**, **cushioning materials** with special requirements, as well as **tapestry**, carpets and other **decorations**.

Fig 2.18 shows some wool products.

(a) Wool gloves（羊毛手套） (b) Wool shoe wipe（羊毛鞋擦） (c) Wool waist supportl（羊毛护腰） (d) Handmade woollen carpet （手工羊毛地毯）

(e) Australian wool car cushion （澳毛汽车坐垫） (f) Woollen socks （羊毛袜） (g) Cardigan （羊毛衫） (h) Woollen coat （羊毛大衣） (i) Woollen sweater set （羊毛衫套装）

Fig 2.18

（j）Wool scarf（羊毛围巾） （k）Wool cap（羊毛帽） （l）Cashmere car steering wheel cover（羊绒汽车方向盘） （m）Wool casual suit（羊毛休闲西服） （n）Industrial felt（工业毛毡）

Fig 2.18　Some wool products（羊毛产品）

【Text】

蚕丝纤维来源于蚕吐出的长丝，也是一种天然蛋白质纤维[1]。蚕丝有较好的强伸性，纤维细而柔软，富有弹性，吸湿性好。蚕分家蚕和野蚕两大类。家蚕即桑蚕，它的结茧是生丝的原料[2]。野蚕有柞蚕、蓖麻蚕等，柞蚕的结茧可用于缫丝，其他野蚕的结茧不易缫丝，一般将它们切成短纤维作绢纺原料或制成丝绵[3]。

蚕丝是用于纺织品生产的天然纤维中强力最高的纤维之一。蚕丝来自构成蚕茧的纤维[4]，形成蚕茧的长丝截面是三角形的，由两根丝素蛋白纤维和起支撑作用的丝胶基质组成[5]。丝胶占茧丝的20%~30%，作为支撑材料将茧子粘在一起，并在纺纱过程中充当润滑剂。丝素蛋白的平均直径是10~18μm，它是一种蛋白质，含有大量非极性氨基酸，如氨基乙酸和丙氨酸[6]。而丝胶是一种水溶性蛋白质，由丝氨酸和天冬氨酸这类蛋白质组成。覆盖在长丝外的丝胶具有很好的坚韧性，为了制得生丝，需要高温下碱处理才能完全除去[7]。蚕丝是纤维皇后，所以蚕丝被用来生产高档织物。

【Key Words】

be derived from　源自于
filament ['filəmənt]　长丝
eject [i'dʒekt]　喷出，吐出
silkworm ['silkwə:m]　蚕
tensile strength and elongation　强伸性
bombyx mori　家蚕
wild silkworm　野蚕
mulberry silkworm　桑蚕
cocoon [kə'ku:n]　茧，蚕茧
raw silk　生丝
tussah　柞蚕
samia cynthia ricini　蓖麻蚕

reeling　缫丝
silk spinning　绢纺
silk wadding　丝绵
silk worm　蚕蛹
filament ['filəmənt]　长丝
triangular [trai'æŋɡjələ(r)]　三角（形）的
fibroin ['faibrəuin]　丝素，蚕丝蛋白
sericin ['seərisin]　丝胶
lubricant ['lu:brikənt]　润滑剂，润滑油
abound in　富含
non-polar ['nɔnp'əulər]　非极性的
amino [ə'mi:nəu] acids　氨基酸类，氨

基酸
glycine ['glaisi:n] 甘氨酸,氨基乙酸
alanine ['æləni:n] 丙氨酸
water-soluble ['wɔ:tərsɔlj'ubl] 可溶于水的
aspartic [ə'spa:tik] acid 天(门)冬氨酸
enwarp 包含
queen [kwi:n] 女王,王后
luxury ['lʌkʃəri] 奢侈,豪华

【Key Sentences】

1. **Silk** fiber **is derived from** the **filaments ejected** by **silkworms** and is also a kind of natural protein fiber.

2. The **bombyx mori** is **mulberry silkworm**, and its **cocoon** is the raw material of **raw silk**. Fig 2.19 shows the cocoon.

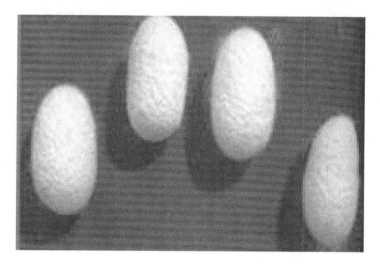

Fig 2.19 Cocoon(蚕茧)

3. The **tussah**'s cocoon can be used for **silk reeling**, but other wild silkworm's cocoons are not easy to reel silk, which are generally cut into short fibers for the materils of **silk spinning** or making **silk wadding**.

4. Silk originates from the fiber used to produce the **cocoon** of the **silk worm**.

Fig 2.20 shows the silk worm.

5. The **filaments** of silk that form the cocoon are **triangular** in shape and are composed of two fibers of **fibroin** and supporting **matrix** of **sericin**.

Fig 2.21 shows the cross section of silk.

6. Fibroin has an average diameter of about 10 to 18 microns and is a protein that **abounds in non-polar amino acids** such as **glycine** and **alanine**.

7. The sericin that **enwarps** the filaments is highly tenacious and requires high temperature alkaline processing to remove it completely in order to obtain raw silk.

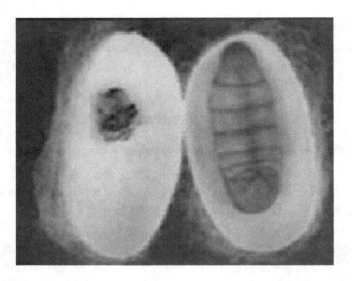

Fig 2.20　Silk worm(蚕蛹)

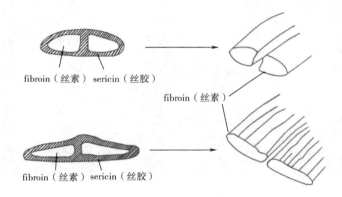

Fig 2.21　Cross section of silk(蚕丝横截面)

【Text】

　　蚕丝湿润时具有较低的湿弹性和折皱,在护理可洗涤的丝绸时要注意这一点[1]。蚕丝有一定程度的弹性回复。蚕丝具有温暖、干燥的接触面,是很好的绝缘体,很少用它作为织物来保暖。蚕丝吸湿性好,干强大,但湿润时会失去强度[2]。长时间暴露在阳光下会使丝绸强力变差,更严重的是,汗水对丝绸服装会产生不利的影响[3]。蛋白质对酸不敏感,但对碱敏感[4]。

　　丝织物可以轻薄(似纱),也可厚实丰满[5]。丝织物不仅用于衣着,还可作日用及装饰品,在工业及国防上也有重要用途[6]。柞蚕丝具有坚牢、耐晒、富有弹性、滑挺等优点,在我国丝绸产品中占有相当的地位[7]。

【Key Words】

resiliency ［ri'ziliənsi］ 弹性　　　　　　　　elastic recovery　弹性回复

insulator ['insjuleitə(r)] 绝缘,隔热
weaken ['wiːkən] 衰减,(使)削弱
exposure [ik'spəuʒə(r)] 暴露
sunlight 阳光
sensitive ['sensətiv] 敏感的
alkali ['ælkəlai] 碱
plump [plʌmp] 丰满的

daily use 日用
national defense 国防
firmness 坚牢
lightfastness 耐光
considerable [kən'sidərəbl] 相当多(或大、重要等)的

【Key Sentences】

1. Silk haslow wet **resiliency** and **wrinkles** easily when it is wet, so we should watch for this when caring for washable silk.

2. Silk absorbs moisture well, and is strong when dry although it does lose strength when wet.

3. Silk can be **weakened** by long **exposure** to **sunlight**, more seriously, perspiration has adverse effects on silk clothing.

4. Proteins are not **sensitive** to acid but sensitive to **alkali**.

5. Silk fabrics can be light and thin like yarn, or thick and **plump**.

6. Silk fabrics are not only used for clothing, but also for **daily use** and decoration, and they also have important applications in industry and **national defense**.

7. Tussah silk has the advantages of firmness, **lightfastness**, elasticity and smoothness, and it occupies a **considerable** position in China's silk products.

Fig 2.22 shows some silk fabrics.

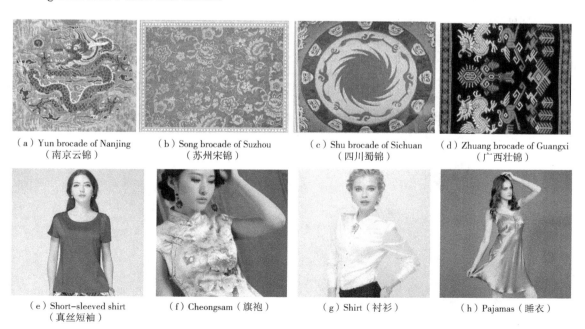

(a) Yun brocade of Nanjing （南京云锦）　　(b) Song brocade of Suzhou （苏州宋锦）　　(c) Shu brocade of Sichuan （四川蜀锦）　　(d) Zhuang brocade of Guangxi （广西壮锦）

(e) Short-sleeved shirt （真丝短袖）　　(f) Cheongsam（旗袍）　　(g) Shirt（衬衫）　　(h) Pajamas（睡衣）

Fig 2.22

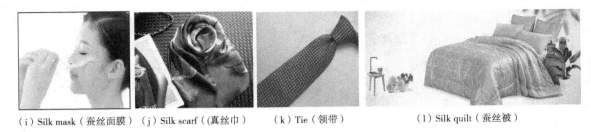

(i) Silk mask(蚕丝面膜)　(j) Silk scarf((真丝巾)　(k) Tie(领带)　(l) Silk quilt(蚕丝被)

Fig 2.22　Some silk fabrics(真丝织物)

Part 2　*Trying*

1. Quickly distinguish wool fiber and silk fiber by visual inspection and touch method.

| Method | wool fiber | silk fiber |

2. Observing the difference between wool fiber and silk fiber in the longitudinal and transverse directions by electron Fiber Mirror.

| Method | wool fiber | silk fiber |

3. Describe the burning characteristics of wool fiber and silk fiber (including odour, ash, etc.) by combustion method.

Method	wool fiber	silk fiber

Part 3 *Thinking*

1. What wool and silk fabrics have you seen in your daily life? What impressed you with their wearing performance?

_____
_____
_____

2. Go to clothing and home textile stores near the school, and find wool and silk fabrics.

_____
_____

3. Comprehensively compare the price differences between cotton, hemp, and silk textiles, and talk about the textiles they are suitable for processing.

_____
_____
_____

Task Three Chemical Fibers(化学纤维)

Part 1 *Learning*

【Text】

化学纤维的制造方法是从蚕吐丝过程中得到的启示[1]。化学纤维的制造,首先,需要成纤高聚物(能制造纤维的高分子化合物)[2]。其次,需要将成纤高聚物制备成纺丝液。若成纤高聚物分解温度高于熔点可通过加热直接熔化成熔体纺丝液,这种方法称为熔融法[3];若成纤高聚物的分解温度低于熔点,可选择适当的溶剂将其溶解制成纺丝溶液,这种方法称为溶液法。熔融法制备的纺丝液采用熔体纺丝工艺制造化学纤维;溶液法制备的纺丝液采用溶液纺丝工艺制造化学纤维,根据去除溶剂的不同方法,将溶液纺丝工艺分为干法纺丝技术与湿法纺丝技术[4]。

化学纤维可分为两类:通过天然聚合物的转化制成的再生纤维和由合成聚合物制成的合成纤维[5]。如果聚合物形成,则化学原料(纺丝)转化成流体状态的液体(纺丝液)并被迫通过喷丝头[6]。挤出孔喷丝头除圆形之外,其他形状也可用于一些纤维的成型。再生纤维可以细分为两种类型:再生纤维素纤维和再生蛋白质纤维[7]。

【Key Words】

chemical ['kemikl] fiber　化学纤维
inspiration [ˌinspə'reiʃn]　灵感,启发灵感的人或物
fiber-forming polymer　成纤高聚物
macromolecular compound　高分子化合物,大分子化合物
fusant　熔体
spinning solution　纺丝液
solvent ['sɔlvənt]　溶剂
dissolve…into　把…溶解成
fusant spinning process　熔体纺丝工艺
solution spinning process　溶液纺丝工艺
be classified ['klæsəˌfaid] into　分(类)为…
category ['kætəgəri]　种类,类别
regenerated [ri'dʒenəreitid] fiber　再生纤维
transformation [ˌtrænsfə'meiʃn]　转变,转化
synthetic [sin'θetik] fiber　合成纤维
convert [kən'və:t]　转变
fluid ['flu:id]　流体的,液体的
state [steit]　状况,状态
spinneret ['spinəˌret]　喷丝头
extrusion [ik'stru:ʒn]　挤出,推出
sub-classify　再分类,次分类

【Key Sentences】

1. The manufacturing method of **chemical fiber** is an **inspiration** from the process about silk ejecting of silkworm.

2. Firstly, the manufacture of chemical fibers requires the **fiber-forming polymer** (**macromolecular compound** capable of producing fibers).

3. If the decomposition temperature of the fiber-forming polymer is higher than the melting point, it can be directly melted into the **fusant spinning solution** by heating, and this method is called the melting method.

Fig 2.23 shows the fusant spinning process.

4. The spinning solution prepared by the solution method uses the **solution spinning process** to produce chemical fibers, and the solution spinning process is divided into **dry spinning technique** and **wet spinning technique** according to different methods for removing the solvent.

Fig 2.24 shows the solution spinning process.

5. **Chemical fibers** can **be classified into** two **categories**: **regenerated fibers** made by **trans-**

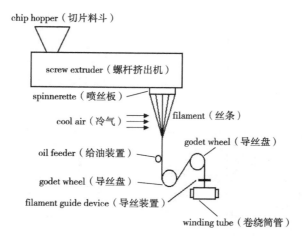

Fig 2.23　Fusant spinning process(熔体纺丝工艺)

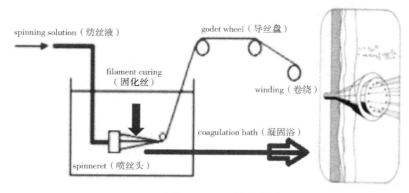

(a) Wet spinning (湿法纺丝)

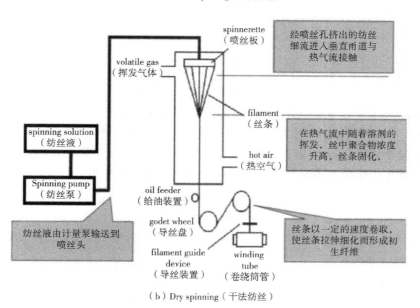

(b) Dry spinning (干法纺丝)

Fig 2.24　Solution spinning process(溶液纺丝工艺)

formation of natural polymers and **synthetic fibers** made by synthetic polymers.

6. If the polymers are formed, the chemicals are **converted** into a liquid of **fluid state** and forced through a **spinneret**.

Fig 2.25 shows the spinneret of spinning about chemical fiber.

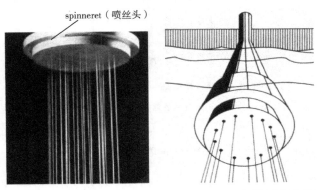

Fig 2.25   Spinneret of spinning about chemical fiber(化学纤维的喷丝头)

7. The regenerated fibers can be **sub-classified** into two types: regenerated cellulosic fibers and regenerated protein fibers.

Tab 2.1 shows the diagram about classification of chemical fibers.

Tab 2.1   classification of chemical fibers(化学纤维的分类)

```
                        ┌─ regenerated fiber ─┬─ regenerated cellulosic fibers(再生纤维素纤维)
Chemical fiber ────────┤   (再生纤维)         └─ regenerated protein fibers(再生蛋白质纤维)
(化学纤维)              │
                        └─ synthetic fiber(合成纤维)
```

【Text】

在使用天然聚合物作为原料的再生纤维中,有一些是次要的,例如,蛋白质或甲壳质的纤维[1],大多数讨论的纤维是再生纤维素纤维(CMF),它使用纤维素作为原料。可以注意到,直到1953年在美国,所有的再生纤维素纤维都被归类为人造丝(黏胶)[2]。在20世纪90年代,一种新型的重组纤维素纤维Lyocell以商业方式引入。Lyocell通过从溶剂中重构纤维素,使用化学制品与加工方法生产出优质的再生纤维素纤维,对环境的影响(污染)很小[3]。

【Key Words】

chitin [ˈkaitin] 甲壳质,甲壳素
rayon [ˈreiɒn] 黏胶纤维
reconstitute [ˌriːˈkɔnstitjuːt] 再组成,再构成
Lyocell 莱赛尔纤维,木浆纤维

commercially [kəˈməːʃəli] 商业上,通商上
solvent [ˈsɔlvənt] 溶剂
superior [suːˈpiəriə(r)] 在(质量等方面)较好的

impact [ˈimpækt] 影响

【Key Sentences】

1. Of those regenerated fibers using natural polymers as raw materials, some are minor, such as fibers from protein or **chitin**.

2. **It can be noted that until** 1953 **in the United States all CMF fibers were classified as rayon**.

3. Lyocell is produced by reconstituting cellulose from **solvent**, using chemicals and processing methods to produce **superior** regenerated cellulosic fibers, which has little **impact** on the **environment**.

【Text】

黏胶纤维是历史最悠久、最早的化学纤维,占世界再生纤维素纤维生产的最大部分[1]。与大多数化学纤维不同,黏胶纤维不是人工合成的。黏胶纤维的原料来源广泛,成本低廉,从不能直接用于纺织加工的纤维素原料,如棉短绒、木材、芦苇等中提取纯净纤维素[2]。也就是说,黏胶纤维由木浆和纤维素基原料制成。

黏胶纤维的重要性在于其多功能性,实际上它是第一种能够加工生产的纤维。黏胶纤维的特性更类似于天然纤维素纤维,如棉或亚麻。有规则再生丝的长度或纵向外观呈现均匀的直径和相互平行的凹条纹线[3]。黏胶纤维由湿法纺丝制成,其横截面显示出高度不规则或锯齿状边缘,纵向条纹是由不规则横截面形成的[4]。

【Key Words】

linter [ˈlintə] 短绒
reed [riːd] 芦苇
that is to say 也就是说
cellulose-based 纤维素基
account for (在数量、比例上)占
viscose fiber = rayon 黏胶
wood pulp [pʌlp] 木浆
lie in [lai] 在于
versatility [ˌvəːsəˈtiləti] 多用途的
viable [ˈvaiəbl] 切实可行的
be similar to 与……相似

linen [ˈlinin] 亚麻布
longitudinal [ˌlɔŋgiˈtjuːdinl] 纵向的
exhibit [igˈzibit] 呈现
uniform [ˈjuːnifɔːm] (形状、性质等)一样的,规格一致的
diameter [daiˈæmitə(r)] 直径
interior [inˈtiəriə(r)] 内部的
parallel [ˈpærəlel] 平行的
serrated 边缘呈锯齿状的,有锯齿形边缘的

【Key Sentences】

1. **Rayon**, the oldest and the first MF has **accounted for** the greatest part of world CMF fiber production.

2. **Viscose fiber** has a wide range of raw materials and its cost is low, because it can extract pure

cellulose from raw materials such as cotton **linters**, wood and **reed**, which cannot be directly used in textile processing.

3. The length or **longitudinal** appearance of regular rayon **exhibits uniform diameter** and **interior parallel** lines called striations.

4. The cross section of the viscose fiber shows highly irregular or **serrated** edges. Those striations are the results of the irregular cross section.

Fig 2.26 shows the longitudinal appearance and cross section of rayon.

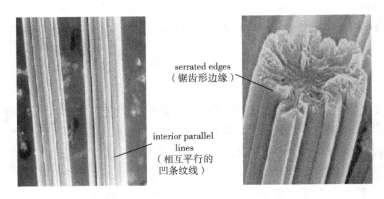

Fig 2.26 Longitudinal morphology and cross section of rayon(黏胶纤维的纵向形态和横截面)

【Text】

黏胶纤维具有丝绸般的美学特性以及极好的悬垂性,并保留了其丰富的绚丽色彩[1]。黏胶纤维的吸湿性是化学纤维中最好的,标准回潮率约为13%。黏胶纤维的染色性能良好,染色色谱全,染色牢度好[2]。黏胶纤维的强度较低,湿强为干强的40%~50%。与棉和聚酯纤维相比,黏胶纤维具有舒适、防潮、柔软和抗起球等性能优点[3]。

黏胶纤维根据纺丝工艺的不同,可分为普通黏胶纤维、高湿模量黏胶纤维(富强黏胶纤维)和强力黏胶纤维[4]。普通黏胶纤维具有明显的皮芯结构。普通黏胶短纤可与其他纤维混纺成纱,用于各类纺织品[5];普通黏胶长丝可以纯织,也可与蚕丝、棉纱、合成纤维长丝等交织,用于制作服装面料、床上用品及装饰品等[6]。高湿模量黏胶纤维具有较高的聚合度、强力和湿模量,强度高于普通黏胶纤维;高湿模量黏胶纤维若在湿态下,单位线密度可承受22.0cN的负荷,且在此负荷下的湿伸长率不超过15%,为富强黏胶纤维[7];富强黏胶纤维横截面近似圆形,结构近乎全芯层,其耐碱性好,织成的织物挺括,洗涤后不会收缩和变形,较为耐穿。强力黏胶丝结构为全皮层,是一种高强度、耐疲劳性能良好的黏胶纤维,强度可达棉的两倍以上,广泛用于工业生产,可做汽车轮胎帘子布,也可以制作运输带、帆布等。

【Key Words】

aesthetic [iːsˈθetik] 审美的, 美的, 美学的

superb [suːˈpəːb] 极佳的,质量极高的

standard moisture regain 标准回潮率

color fastness 染色牢度
color spectrum 色谱
drape[dreip] (帘、幕、衣、裙等)悬挂状
rich 富有的,丰富多彩的
brilliant ['briliənt] 美好的,闪耀的
characteristic [ˌkærəktə'ristik] 特性,特征
softness ['sɔftnəs] 柔软
antipilling 抗起毛性
ordinary viscose fiber 普通黏胶纤维
high wet modulus viscose fiber 高湿模量黏胶
strong viscose fiber 富强黏胶纤维
high tenacity viscose fiber 强力黏胶纤维
sheath-core structure 皮芯结构
interweave 交织
polymerization degree 聚合度
all-core 全芯层
alkali resistance 耐碱性
crisp 挺括
shrink 收缩
deform 变形
all-skin 全皮层
fatigue resistance 耐疲劳性

【Key Sentences】

1. Rayon has a silk-like **aesthetic** property with **superb drape** and retain its **rich brilliant** colors.

2. The viscose fiber has good dyeing performance, complete dyeing **color spectrum**, bright color and good **color fastness**.

3. Compared with cotton and polyester, rayon has advantages of **characteristics**, such as comfort, moisture, **softness**, and **antipilling**.

4. According to the different spinning processes, viscose fiber can be divided into **ordinary viscose fiber**, **high wet modulus viscose fiber** (**strong viscose fiber**) and **high tenacity viscose fiber**.

Fig 2.27 shows the cross section of different kinds of viscose fibers.

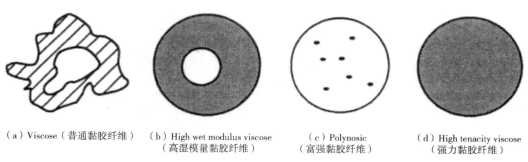

(a) Viscose (普通黏胶纤维)　　(b) High wet modulus viscose (高湿模量黏胶纤维)　　(c) Polynosic (富强黏胶纤维)　　(d) High tenacity viscose (强力黏胶纤维)

Fig 2.27 Cross section of different kinds of viscose fibers(不同黏胶纤维的横截面)

5. Ordinary viscose staple fiber can be blended with other fibers into yarn for various textiles.

6. Ordinary viscose filaments can be purely woven, or be **interwoven** with silk, cotton yarn and synthetic fiber for making apparel fabrics, beddings and decorations.

7. The high wet modulus viscose fiber is the strong viscose fiber if it can withstand the load of 22.0 cN per unit linear density under wet and its wet elongation under this load does not exceed 15%.

【Text】

莱赛尔(Lyocell)纤维是通过将从木浆中获得的纯纤维素溶解在氧化胺中制成的[1]。Lyocell 的加工过程基本无污染,估计只需要黏胶加工所需时间的六分之一。Lyocell 具有与棉和一些聚酯纤维相当的干强度,非常高的湿模量,并且在潮湿时保持85%的干强度,因此,它可以很好地适用于家庭洗涤[2]。当然,Lyocell 由纤维素组成,具有良好的吸湿性、抗静电性,手感凉爽,耐热性好的优点[3]。Lyocell 的商品名为天丝(Tencel)。

【Key Words】

amine oxide ['æmi:n 'ɔksaid] 氧化胺
pollution-free [pə'lu:ʃnfr'i:] 无污染
comparable ['kɔmpərəbl] 可比较的,比得上的
modulus ['mɔdjuləs] 模量
endurable 经久耐用
laundering ['lɔ:ndəriŋ] 洗涤(衣等),洗烫(衣等)

【Key Sentences】

1.Lyocell is made by **dissolving** purified cellulose from wood pulp in an **amine oxide**. Fig 2.28 shows the Lyocell.

Fig 2.28 Lyocell(莱赛尔)

2.Lyocell has dry strength **comparable** to cotton and some polyester, very high wet **modulus**, and it retains 85 percent of its dry strength when wet, so it will **stand up** well to home **laundering**.

3. Of course, being made up of cellulose, it has the andvantages of good moisture absorption, good static resistance, cool hand, and good heat resistance.

Fig 2.29 shows the trade name of Lyocell.

Fig 2.29　Trade name of Lyocell(莱赛尔的商品名)

【Text】

莫代尔(Modal)纤维是一种新型的再生纤维素纤维,和 Lyocell 一样,纤维的生产加工过程清洁无毒。莫代尔纤维采用欧洲的榉木,经打浆和纺丝而成,因此,原料是 100%天然的,对人体无害[1];另外,它能自然分解,对环境无害,被称为绿色纤维,是 21 世纪最具有潜质的纤维[2]。莫代尔纤维吸水、透气性能都优于棉,具有较高的上染率,莫代尔纤维可与其他纤维混纺制成混纺纱,使织物具有丝绸般的光泽,悬垂性好,手感柔软、滑爽,有极好的尺寸稳定性和耐穿性,是制作高档服装、流行时装的首选面料。莫代尔纤维针织面料是目前国内外市场上颇为紧俏的服装面料之一[3]。

竹浆纤维也是一种再生纤维素纤维。它是先将竹子制成适合纺丝的竹浆粕,然后经湿法纺丝获得的[4]。竹浆纤维强力好,耐磨性、吸湿性、悬垂性好,有丝质感觉,手感柔和光滑。竹浆纤维纵向表面具有光滑、均一的特征,但有多条较浅的沟槽,而横截面边缘为不规则锯齿形[5]。竹浆纤维具有较好的抱合力,成纱性能好。竹浆纤维的标准回潮率可达 12%,染色性好且不易褪色,而且竹浆纤维具有天然抗菌性能。由于竹浆纤维比其他纤维更具有吸湿快干性能,因此,适合做夏季服装、运动服和贴身衣物。

【Key Words】

non-toxic　无毒的
beech wood　榉木
beating　打浆
promising　有希望的,有前途的
dyeing rate　上染率
drapability　悬垂性
dimensional stability　尺寸稳定性
durability　耐穿性
high-end clothing　高档服装

fashionable dress　流行时装
first choice　第一选择
hard-to-get　紧俏的,广受欢迎的
bamboo pulp fiber　竹浆纤维
a plurality [pluəˈræləti] of　多个,数个
shallow　浅的
groove　沟槽
zigzag [ˈzigzæg]　锯齿形状
fade [feid]　褪色

【Key Sentences】

1. Modal fibers are made from European **beech wood** by beating and spinning, so the raw material is 100% natural and harmless to the human body.

Fig 2.30 shows the longitudinal appearance and cross section of modal fiber. Fig 2.31 shows the modal fiber.

2. Modal fiber can be naturally decomposed and harmless to the environment, therefor, it is called green fiber and is the most **promising** fiber in the 21st century.

3. Modal knitted fabric is one of the underwear fabrics which are the most **hard-to-get** fabrics on the domestic and foreign markets.

4. **Bamboo pulp fiber** is made by first making bamboo into the bamboo pulp suitable for spinning, then obtained by wet spinning.

Fig 2.32 shows the longitudinal morphology and cross section of bamboo pulp fiber.

5. The longitudinal surface of the bamboo pulp fiber has a smooth and uniform characteristic, but also has **a plurality of shallow grooves**, while the cross-sectional edge of it has an irregular **zigzag**.

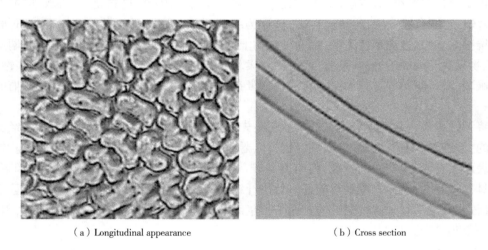

(a) Longitudinal appearance　　　　　　(b) Cross section

Fig 2.30　Longitudinal morphology and cross section of modal fiber(莫代尔纤维纵向形态和横截面)

【Text】

大豆蛋白纤维是一种再生植物蛋白纤维,是从豆渣中提取球蛋白经化学纺丝加工而成[1]。大豆蛋白纤维吸湿性好,力学性能好,常规洗涤下不收缩,抗皱性好,易洗快干。因大豆蛋白纤维含有多种人体所必需的氨基酸,所以具有保健功能。大豆蛋白纤维面料具有羊绒般的手感、蚕丝般的光泽、羊毛般的保暖性,且吸湿透气,悬垂性好,可做高档衬衫、内衣。

牛奶蛋白纤维是牛奶中分离和提纯出来的蛋白质与聚乙烯醇缩甲醛聚合接枝而成的新型化学纤维[2]。牛奶蛋白纤维属于再生蛋白质纤维。牛奶蛋白纤维中也含有人体所需的多种氨基酸,因此,牛奶蛋白纤维面料常用于贴身穿着,有润肤养肤的功效。此外,牛奶蛋白纤维织物具有质地轻盈、柔软、透气、导湿、爽身的优点,是制作儿童服饰和女士内衣的理想面料[3]。

Fig 2.31　Modal fiber(莫代尔纤维)

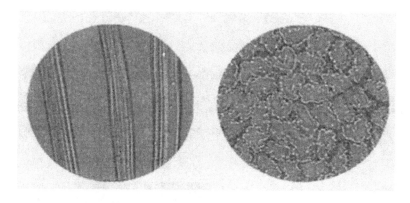

Fig 2.32　Longitudinal morphology and cross section of bamboo pulp fiber(竹浆纤维的纵向形态和横截面)

【Key Words】

soy [sɔi] protein fiber　大豆蛋白纤维
vegetable protein　植物蛋白
globulin ['glɔbjuin]　球蛋白
bean dreg　豆渣
wrinkle resistance　抗皱性
quick-drying washable　易洗快干
cashmere ['kæʃmiə(r)]　羊绒
milk protein fiber　牛奶蛋白纤维
polymerization　聚合

grafting　接枝
polyethylene [ˌpɔli'eθəliːn] alcohol
formaldehyde [fɔː'mældihaid]　聚乙烯醇缩甲醛
purify　提纯
moisturize ['mɔistʃəraiz]　使皮肤湿润,滋润
nourish ['nʌriʃ]　滋养,养育
non-cling [kliŋ]　爽身

【Key Sentences】

1. **Soy protein fiber** is a kind of regenerated **vegetable protein** fiber, which is obtained by chemical spinning of **globulin** extracted from **bean dregs**.

Fig 2.33 shows the soy protein fiber.

Fig 2.33　Soy protein fiber(大豆蛋白纤维)

2. Milk protein fiber is a new type of chemical fiber formed by the **polymerization** and **grafting** between **polyethylene alcohol formaldehyde** and the protein separated and **purified** from milk.

Fig 2.34 shows the milk protein fiber.

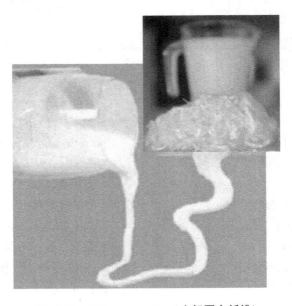

Fig 2.34　Milk protein fiber(牛奶蛋白纤维)

3. In addition, milk protein fiber fabrics which have the advantages of light weight, softness, breathability, moisture permeability and **non-cling** make them ideal for children's clothing and women's underwear.

【Text】

纺织工业所用的合成纤维通常是由石油、煤、天然气加工的[1]。这些原料中的单体通过各种化学反应聚合成高分子聚合物。尼龙是世界上出现的第一种合成纤维,尼龙是聚酰胺纤维(锦纶)的一种说法[2]。锦纶于1938年首次作为刷毛和鱼线销售,并在20世纪40年代初作为尼龙袜或"尼龙"销售。锦纶长丝光滑有光泽。从横截面看,锦纶通常是完美的圆形,特殊的为三叶形锦纶,如安特龙、卡登、库穆洛夫特和奎阿纳[3]。纵向放大显示锦纶直径均匀且相对透明,并具有略微斑点的外观[4]。

【Key Words】

petroleum [pə'trəuliəm] 石油
coal [kəul] 煤
natural gas 天然
monomers 单体
polymerize ['pɔliməraiz] (使)聚合
molecular [mə'lekjələ(r)] 分子的,分子组成的
nylon 锦纶(尼龙)
term [tə:m] 术语
polyamide [pɔli'æmaid] 聚酰胺
bristle ['brisl] 刷毛
fish line 鱼线
hosiery ['həuziəri] 袜类,针织内衣
shiny ['ʃaini] 发光的,光亮的
round [raund] 圆形的,弧形的
trilobal [trai'ləubəl] 三叶形的
magnification [ˌmægnifi'keiʃn] 放大,夸大
transparent [træns'pærənt] 透明的
speckled ['spekld] 有斑点的

【Key Sentences】

1. Synthetic fibers used in textiles are generally made from **petroleum**, **coal** or **natural gas**.
2. **Nylon** is the first synthetic fiber in the world, and it is a **term** for **polyamide** fiber.
3. When viewed in cross section, nylon is usually perfectly **round**, exception to this are **trilobal** nylons such as Antron, Cadon, Cumuloft, and Qiana.
4. Longitudinal **magnification** shows relatively **transparent** fibers of uniform diameter, with a slightly **speckled** appearance.

Fig 2.35 shows the longitudinal enlargement of nylon.

【Text】

锦纶(尼龙)的机械强度高,韧性好,有较高的抗拉、抗压强度[1]。耐疲劳性能突出,锦纶制件经多次反复屈折仍能保持原有机械强度[2]。锦纶的摩擦系数小,耐磨性是常见合成纤维中最好的[3]。锦纶的耐热性和耐光性都较差,耐碱性优良,但耐酸性较差。

锦纶由熔体纺丝法制得,其吸湿性是合成纤维中较好的,在一般大气条件下,回潮率可达

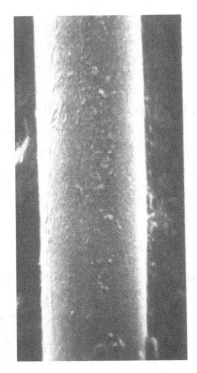

Fig 2.35　Longitudinal enlargement of nylon（锦纶的纵向放大图）

4.5%~7%。锦纶较易染色,弹性优良,伸长率在6%以下时,弹性回复率接近100%[4]。锦纶在小负荷下容易变形,其初始模量在常见纤维中是最低的,因此,手感柔软,但锦纶织物的保形性和硬挺性不及涤纶织物[5]。

锦纶的主要应用为长丝,可用于织制袜子、围巾、衣料及用作牙刷鬃丝等,还可以织制地毯;在工业上,可应用于轮胎帘子线、绳索、渔网的制作;国防上,用锦纶织制降落伞等。锦纶短纤维可与棉、毛和黏胶纤维混纺,其混纺织物具有良好的耐磨性和强度,在服装面料上应用较多。

【Key Words】

toughness [tʌfnəs] 韧性;坚韧
compressive [kəmˈpresiv] strength 压缩强度
fatigue [fəˈtiːg] resistance [riˈzistəns] 抗疲劳性,耐疲劳性
outstanding [autˈstændiŋ] 显著的,凸出的
coefficient [ˌkəuiˈfiʃnt] friction 摩擦系数
wear resistance [weə riˈzistəns] 耐磨性
light resistance 耐晒性
heat resistance 耐热性

light resistance 耐光性
alkali resistance 耐碱性
acid resistance 耐酸性
atmospheric condition 大气条件
shape retention 保形性
stiffness 刚性
be inferior to 差于,劣于
toothbrush [ˈtuːθbrʌʃ] 牙刷
fishing net 渔网
parachutes [ˈpærəʃuːts] 降落伞

【Key Sentences】

1. Nylon has high mechanical strength, good **toughness** and high tensile and **compressive** strength.

2. The **fatigue resistance** is **outstanding**, and the nylon fiber products can maintain the original mechanical strength after repeated bending.

3. Nylon has a small **coefficient friction**, and its **wear resistance** is the best in common synthetic fibers.

4. The nylon has excellent elasticity and the elastic recovery rate is close to 100% when the elongation is below 6%.

5. Nylon is easily deformed under small load, and its initial modulus is the lowest among common fibers. Therefore, it has soft handle, but the **shape retention** and **stiffness** of the nylon fabric are **inferior** to those of polyester.

【Text】

聚酯纤维是1941年开发的第二种合成纤维类,由于其性能的平衡,是迄今为止使用最广泛的一种[1]。与锦纶相比,它具有极大的多功能性和较便宜的制造工艺[2]。聚酯的纵向视图显示出均匀的直径、光滑的表面和棒状外观;普通聚酯的横截面是圆形的[3]。横截面形状有几种变化,最常见的包括Dacron(品牌名)的三叶变体,Fortrel(品牌名)的T形变体,Hoeschst(Trevira Star)(品牌名)的五角形变体,Encron(品牌名)的八边形变体,以及Kodel(品牌名)的"三边"变体。

涤纶吸湿性差,在一般大气条件下,回潮率只有0.4%左右,因而纯涤纶织物穿着时有闷热感,但易洗快干,具有"洗可穿"的美称。涤纶大分子不含亲水基团,分子间隙小,因此,染色较难。涤纶的断裂强度和断裂伸长率均大于棉纤维,涤纶的初始模量较高,所以,涤纶织物挺括、抗皱,尺寸稳定,保形性好;涤纶对酸较稳定,但只耐弱碱;涤纶的耐热性优良;涤纶因吸湿性差,比电阻高,可作为优良的绝缘材料,但易积聚电荷产生静电,吸附灰尘;涤纶的耐光性仅次于腈纶。

【Key Words】

rodlike 棒状
trilobal [traiˈləubəl] 三叶形的
variant [ˈveəriənt] 变体
pentalobal 五叶形
eight-sided-star 八边形
trilateral [ˌtraiˈlætərəl] 三边形
sultry [ˈsʌltri] 闷热的
reputation [ˌrepjuˈteiʃn] 美称,名声

hydrophilic group 亲水基团
intermolecular gap 分子间隙
specific resistance 比电阻
insulating material 绝缘材料
accumulate [əˈkju:mjəleit] 积累,积聚
electric charge 电荷
generate [ˈdʒenəreit] 产生
static electricity 静电

【Key Sentences】

1. Polyester was the second synthetic fiber type to be developed(1941) and is now by far the most widely used, due to a balance of properties.

2. Polyester has great versatility and less expensive manufacturing process compared to nylon.

3. The longitudinal view of polyester exhibits uniform diameter, smooth surface, and a **rodlike** appearance. The cross section of regular polyester is round.

Fig 2.36 shows the longitudinal view and cross section of regular polyester.

4. Polyester has poor moisture absorption and high **specific resistance**, so it can be used as the excellent **insulating material**, but it is easy to **accumulate electric charges** to **generate static electricity** and absorb dust.

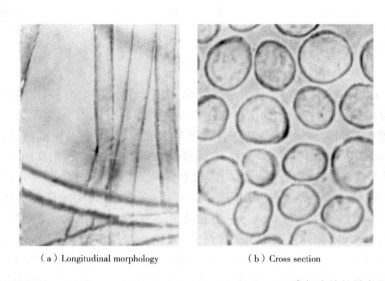

(a) Longitudinal morphology　　　(b) Cross section

Fig 2.36　Longitudinal morphology and cross section of regular polyester(常规涤纶的纵向形态和横截面)

【Text】

从性能的角度来看,聚酯纤维具有与锦纶相近的强度和耐磨性,并且还具有永久热定形——所有"一流"纤维特性[1]。聚酯纤维优于锦纶的特性是模量,这意味着它有弹性并且可以从应变中恢复良好,且抗拉伸性能优于锦纶。聚酯纤维还能比锦纶更好地从皮肤传导水分,干燥更快,在寒冷条件下更柔软[2]。聚酯纤维也有特殊的缺点,如顽固地保持油性污渍;它吸收的水分很少,因此,正常穿戴的舒适性不佳[3];像锦纶一样,由于纤维的强度很高,聚酯织物显示出顽固的起毛起球[4]。

涤纶无论在服装、装饰还是工业中应用都相当广泛。涤纶短纤可与棉、毛、丝、麻或其他化纤混纺,用于服装、装饰等。涤纶长丝可用针织、机织技术制成各种仿真型内外衣[5]。涤纶长丝也因其良好的物理化学性能,广泛用于轮胎帘子线、工业绳索、传动带、滤布、船帆、篷帐等工业制品[6]。随着新技术、新工艺的不断应用,对涤纶进行改性制得了抗静电、抗起毛起球、阳离子可染等涤纶[7]。涤纶以其发展速度快、产量高、应用广,堪称当今化学纤维之冠[8]。

【Key Words】

permanent ['pə:mənənt] 永久(性)的,永恒的
heat set 热定型
ahead of 在…之前
springy ['spriŋi] 有弹性的
strain [strein] 应力,拉力
stretching ['stretʃiŋ] 拉伸,伸长
supple ['sʌpl] 柔软的,易弯曲的
drawback ['drɔ:bæk] 缺点,劣势
oily stains 油渍
tenaciously [tə'neiʃəsli] stubborn
['stʌbən] 顽固的,固执的
simulated ['simjuleitid] 模拟的,仿真的
sail 船帆
cationic-dyeable 阳离子可染的
crown [kraun] 王冠,皇冠,冕

【Key Sentences】

1. Polyester has strength and abrasion next to nylon, and also takes a **permanent heat set**—all "first-class" fiber properties, from the point view of performance.

2. Polyester also conducts moisture away from the skin better than nylon, dries faster, and is more **supple** in cold conditions.

3. Polyester does have particular **drawbacks**, too. For example, it holds **oily stains tenaciously**. It absorbs very little moisture, so it is less comfortable to wear in its standard form.

4. Like nylon, polyester fabrics show **stubborn** pilling because of the great strength of the fiber.

5. Polyester filaments can be knitted and weaved into a variety of **simulated** inner and outer garments.

6. Polyester filaments are also widely used in industrial products such as tire cords, industrial ropes, transmission belts, filter cloths, **sails**, and tents, due to their good physical and chemical properties.

7. With the continuous application of new technologies and new processes, the polyester is modified to obtain antistatic, anti-pilling, **cationic-dyeable** and other polyester.

8. Polyester is the **crown** of today's chemical fiber because of its rapid development, high output and wide application.

【Text】

腈纶是聚丙烯腈在我国的商品名[1],国外则称为奥纶、开司米纶,美国杜邦公司就取名为Orlon。腈纶产量仅次于聚酯和聚酰胺两个合成纤维品种。它柔软、轻盈、保暖、耐腐蚀、耐光似羊毛纤维,密度却比羊毛小,因此,有人造羊毛之称[2]。腈纶虽比羊毛轻10%以上,但强度却大2倍多[3]。

腈纶(PAN)主要由聚丙烯腈组成;腈纶的截面一般为圆形或哑铃形,纵向表面平滑或有1~2根沟槽[4]。腈纶的吸湿性优于涤纶,但比锦纶差,在一般大气条件下,回潮率为2.0%左右,腈纶的强度比涤纶、锦纶低,断裂伸长率与涤纶、锦纶相似,但尺寸稳定性也较差,耐磨性是合成纤维

中较差的。腈纶不会发霉和被虫蛀,且对日光的抵抗性也比羊毛大1倍,比棉花大10倍。腈纶的耐光性是常见纤维中最好的,因此,腈纶织物在篷帐、炮衣、窗帘等应用较多[5]。腈纶可纯纺(如腈纶毛线)或与羊毛混纺制毛线、毛织物等。用腈纶制成的毛线特别是轻软的膨体(毛)纱早就为人们所喜爱[6]。

【Key Words】

acrylic [ə'krilik] fiber 腈纶
trade name 商品名
polyacrylonitrile 聚丙烯腈
orlon 奥纶
cashmilan 开司米纶

DuPont Co. 杜邦公司
artificial wool 人造毛
dumbbell-shape 哑铃形
gun cover 炮衣
bulk(wool)yarn 膨体纱

【Key Sentences】

1. **Acrylic fiber** is the **trade name** of **polyacrylonitrile** in China.

2. Acrylic fiber is soft, light, warm, corrosion-resistant, light-resistant like wool, while its density is smaller than wool, so it is called **artificial wool**.

3. Acrylic fiber is 10% lighter than wool, but its strength is more than 2 times of wool.

4. The cross section of acrylic fiber is generally round or **dumbbell-shape**, and the longitudinal surface is smooth or has 1 or 2 grooves.

Fig 2.37 shows the longitudinal view and cross section of acrylic fiber.

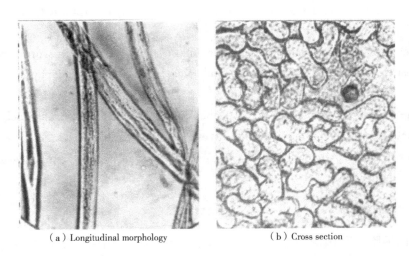

(a) Longitudinal morphology   (b) Cross section

Fig 2.37 Longitudinal morphology and cross section of acrylic fiber(腈纶的纵向形态和横截面)

5. The light resistance of acrylic fiber is the best among common fibers, so there are many applications of acrylic fabrics in tents, **gun covers**, curtains, etc.

6. The yarn made of acrylic fiber, especially light and soft **bulk(wool)yarn**, has long been loved.

【Text】

丙纶是由聚丙烯经熔体纺丝制得的,截面与纵面形态与涤纶、锦纶等相似。大多数聚丙烯的横截面是椭圆形或圆形的,但是它们中有一些具有不规则的近三角形的形状[1]。丙纶的相对密度为0、91g/cm^3,是常用纤维中最轻的纤维[2]。丙纶的弹性好,尺寸稳定性强,回弹性可与锦纶、羊毛相媲美。丙纶耐磨性好,在各类合纤中最不易起球,所以,常用于汽车工业制作沙发布和装饰布等[3]。丙纶几乎不吸湿,回潮率接近零,因此,丙纶具有良好的抗静电性能,不适合细菌和霉菌生存,也不易被虫蛀[4]。丙纶的耐光性较差,易老化。丙纶的化学稳定性优良,耐酸碱能力均较强,并有良好的耐腐蚀性[5]。另外,丙纶很难被沾污,丙纶织物耐脏易洗,能吸收异味。丙纶熔点较低,因此,耐热性能较差[6];丙纶的导热系数在常见纤维中是最低的,因此,保温性能好。由于没有吸湿性,丙纶难以染色;原液染色是使聚丙烯纤维着色的最有效方法[7]。

常规聚丙烯的制造工艺流程短,同时,近几年来技术不断进步,能制成各种适合纺丝的、价廉的聚丙烯切片[8]。丙纶是以常规聚丙烯为原料聚合而成的,所以取源容易。在制取同样厚度、宽度和长度织物条件下,用丙纶纱代替棉纱时,原纱成本可降低很多[9]。丙纶是合成纤维中发展较迟的一个品种,生产以短纤维为主。丙纶短纤维可以纯纺,或与棉、黏胶纤维等混纺,织制服装面料、地毯等装饰用织物、土工布、过滤布、人造草坪等;丙纶做成的纱布不粘伤口,故可用于医疗行业[10];近年来,丙纶还用作土建用布和人工草坪的主要原料。

【Key Words】

polypropylene [ˌpɔliˈprəupəliːn] 聚丙烯,丙纶
olefins [ˈəulifinz] 烯烃
specific gravity 比重
resilience [riˈziliəns] 回弹性,快速回复的能力
pilling 起球
non-hygroscopic 不吸湿的,防潮的
bacteria [bækˈtiəriə] 细菌
mold [məuld] 霉,霉菌
survive [səˈvaiv] 生存,存活
infect 虫蛀

age 老化
chemical stabilit 化学稳定性
acid and alkali resistance 耐酸碱性
corrosion resistance 耐腐蚀性
stain [stein] (被)玷污
odor [ˈəudə] 异味,臭味
coefficient of thermal conductivity 导热系数
heat retention 保温性
inexpensive 廉价的
wound 伤口
solution dyeing 原液染色

【Key Sentences】

1. Most **polypropylene** fibers are elliptical or round in cross section, but some of them have an irregular, almost triangular shape.

2. Polypropylene fiber has a specific gravity of 0.91 g/cm^3 and is the lightest fiber among the commonly used fibers.

Fig 2.38 shows the polypropylene fiber.

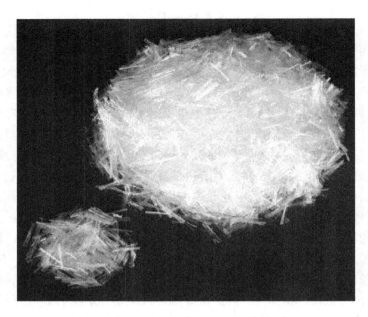

Fig 2.38　Polypropylene fiber(聚丙烯纤维)

3. Polypropylene fiber has good abrasion resistance and is the most difficult to **pilling** in all kinds of synthetic fibers, so it is often used in the automotive industry to make sofa fabrics and decorative fabrics.

4. Polypropylene fiber is almost **non‒hygroscopic**, and the moisture regain is similar to 0, therefore, polypropylene fiber has good antistatic properties, and it is not suitable for **bacteria** and **mold** to **survive** and also not easy to be **infected**.

5. Polypropylene has poor light resistance and is easy to **age**、Polypropylene has excellent **chemical stability**, strong acid and alkali resistance, and good **corrosion resistance**.

6. In addition, polypropylene fiber is difficult to be **stained**, and polypropylene fiber fabric is resistant to dirt and washable, and can absorb **odor**.

7. With almost no moisture absorption, polypropylene fiber is difficult to dye; **solution dyeing** is the most effective way to color the fiber.

8. The conventional polypropylene has a short manufacturing process, and at the same time, the technology has been continuously improved in recent years, and various **inexpensive** polypropylene chips suitable for spinning can be produced.

9. The original yarn cost can be reduced much when polypropylene yarn is used instead of cotton yarn when producing the same thickness, width and length of the fabric.

10. The gauze made of polypropylene fiber is not stick to **wound**, so it can be used in medical industry.

【Text】

主要合成弹性体——氨纶,应仅与橡胶性能进行比较。因为氨纶的强度相对较低,氨纶不具备合成纤维的一般性能[1]。氨纶可以拉伸至原来长度的7倍,当张力释放时立即回复[2]。氨纶通常作为连续长丝出售,尽管已生产一些短纤维用于与其他纤维混合纺纱。在横截面中,莱卡具有圆形或狗骨形状;GLPSPANN 氨纶看起来像熔合的、空心的、轻微锯齿状的细丝[3]。氨纶相当薄弱,强度在 4.4~13.2cN/tex[4],伸长率在 500%~700%,伸长率为2%时的回复率为100%。在50%的伸长率下,氨纶具有99%的回复率;在200%的伸长率下,其回复率为97%。其湿弹性回复与干弹性回复相似。

【Key Words】

elastomer [iˈlæstəmə(r)] 弹性体,人造橡胶
spandex [ˈspændeks] 氨纶
conform to 顺应,符合
instantly [ˈinstəntli] 立即,马上,立刻
lycra 莱卡(氨纶的商品名)
dogbone 狗骨

Glospan Acelon, GlobeMfg 公司生产的一种氨纶纤维品牌
fuse [fju:z] 融合,熔化
hollow-core 空心型芯
serrated [səˈreitid] 边缘呈锯齿状的,有锯齿形边缘的

【Key Sentences】

1. **Spandex** does not **conform to** the general behavior of synthetic fibers, since its strength is relatively low.

Fig 2.39 shows the spandex.

Fig 2.39 Spandex(氨纶)

2. Spandex can stretch up to seven times its original length and recover **instantly** when tension is released.

3. In cross section, **Lycra** has a round or **dogbone** shape, **Glospan** fibers look like **fused**, hol-

low-core, lightly **serrated** filaments.

4. Spandex fibers are also fairly weak, with tenacities ranging from 4.4 to 13.2 cN/tex. The elongation of spandex fibers ranges from 500 to 700 percent, with 100 percent recovery at 2 percent elongation.

【Text】

丙烯酸纤维和聚酯纤维大约同时开发。丙烯酸纤维横截面形状由纺丝方法确定,可能是圆形、狗骨形,或看起来像两根纤维粘在一起[1]。丙烯酸纤维的特性包括与其他再生纤维相似的手感:温暖、干燥和柔软。丙烯酸纤维也以低密度而著名,并具有良好的弹性,这种特性的组合可制成温暖且重量轻的羊毛状织物[2]。消费者使用的所有丙烯酸纤维都是短纤维,并且大部分用于代替羊毛或与羊毛一起混纺,因为丙烯酸纤维比羊毛便宜得多,具有相当好的强度,并且在潮湿时不会失去强度[3]。丙烯酸纤维已经占据了毛毯和中等价格到低价服装的大部分羊毛市场,尤其是婴儿和儿童类的。如上所述,丙烯酸纤维具有相当低的吸收性,具有防污性和可清洁性,但表现出静电聚集和较差的舒适性[4]。丙烯酸纤维比锦纶或聚酯纤维对热更敏感——在滚筒干燥丙烯酸纤维制品时要注意这一点。

【Key Words】

acrylic [əˈkrilik] fiber 丙烯酸纤维
most-like 很像
hand 手感
resiliency [riˈziliənsi] 回弹,弹性
replace 代替
take over 接受,接管,取代

moderate [ˈmɔdərət] 适合的
infant [ˈinfənt] 婴儿,幼儿
cleanability [kliːnəˈbiliti] 可清洁性
tumble [ˈtʌmbl] drying 滚筒干燥
articles 物品,产品

【Key Sentences】

1. The cross-sectional shape of acrylic is determined by the spinning method. They may be round, dogbone, or look like two fibers **stuck** together.

2. Acrylic are also famous for its low density and good **resiliency**—a combination to make into wool-like fabrics that are warm and light.

3. All acrylic in consumer use is staple fiber, and much is used **in place of**, or with wool since acrylic is much less expensive than wool, has fairly good strength, and does not lose strength when wet.

4. Acrylic has quite low absorbency, giving stain resistance and **cleanability**, as noted, but exhibiting static collection and poor comfort in clothing.

Part 2 *Trying*

1. Quickly distinguish **rayon**, **tencel**, **nylon**, **polyester**, **spandex** by visual inspection and touch method.

Project Two Textile Fiber(纺织纤维)

Method　　　　　　　　　rayon　　　　tencel　　　　nylon

Method　　　　　　　　　polyester　　　　spandex

2. Observing the difference of **rayon**, **tencel**, **nylon**, **polyester**, **spandex** in the longitudinal and transverse directions by Electron Fiber Mirror.

Method　　　　　　　　　rayon　　　　tencel　　　　nylon

| Method | polyester | spandex |

3. Describe the burning characteristics of **rayon, tencel, nylon, polyester, spandex** (including odour, ash, etc.) by combustion method.

| Method | rayon | tencel | nylon |

| Method | polyester | spandex |

Part 3 *Thinking*

1. In addition to the chemical fibers mentioned in the text, have you heard of other chemical fibers? What impressed you with their wearing performance?

2. Go to clothing and home textile stores near the school, and find which textile products are made in chemical fiber?

Project Three Yarns and Spinning Technology
（纱线与纺纱技术）

Task One Yarns（纱线）

Part 1 *Learning*

【Text】

纱线是针织物或机织物的基本元素[1]。因此，纱线的纺制几乎与织物的织制一样古老，并且肯定早于有史以来的记载[2]。在史前时期，纤维以简单的方式捻合在一起形成纱线[3]。

【Key Words】

yarn [jaːn] 纱，线
fundamental [ˌfʌndəˈmentl] 基础的，基本的，根本的
element [ˈelimənt] 要素
knitted fabric 针织物
woven fabric 机织物
predate [ˈpriːˈdeit] 居先
prehistory [ˌpriːˈhistri] 史前时期
twist [twist] 加捻，搓
roll [rəul] 滚动，旋转

【Key Sentences】

1. **Yarns** are **fundamental elements** for **knitted** or **woven fabrics**.

Fig 3.1 shows the yarn, woven fabric and knitted fabric.

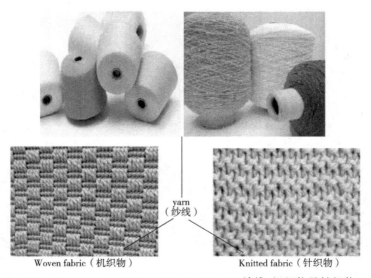

Fig 3.1 Yarn, woven fabric and knitted fabric（纱线，机织物及针织物）

2. Thus, the production of yarn is nearly as old as the manufacture of fabric and **definitely predates** recorded history.

3. In **prehistory** times, fibers were **twisted** together in simple ways to form yarns.

【Text】

通常,纱线可以定义为纤维或长丝的线性组合并形成连续须条[1]。纱线可以由一根或多根连续长丝或许多非连续和相当短的纤维组成。为了克服纤维滑移并形成功能性纱线,短纤维通常具有大量的扭曲或缠结[2]。由短纤维制成的纱线通常被称为短纤纱。可以将两根或更多根单纱捻合在一起以形成合股纱[3]。

【Key Words】

be defined [diˈfaind] as 被定义为…
linear [ˈliniə(r)] 直线的,线形的
assemblage [əˈsemblidʒ] 聚集
filament [ˈfiləmənt] 长丝
continuous [kənˈtinjuəs] 连续的,延伸的
strand [strænd] (绳子的)股
overcome [ˌəuvəˈkʌm] 战胜,克服
slippage [ˈslipidʒ] 滑动

functional [ˈfʌŋkʃənl] 功能的
a great amount of 大量的
entanglement [inˈtæŋglmənt] 纠缠
be referred to as 被称为…
twist [twist] 加捻,搓
ply [plai] yarn = plied [plaid] yarn 合股线

【Key Sentences】

1. Generaly, yarn maybe **defined** as a **linear assemblage** of fibers or **filaments** formed into a **continuous strand.**

2. To **overcome** fiber **slippage** and to form into a **functional** yarn, staple fibers are usually given **a great amount of** twist or **entanglement.**

3. Two or more single yarns can be **twisted** together to form **ply** or **plied yarns**.

Fig 3.2 shows the structure of staple yarn, multifilament, ply yarn, cord and multiply yarn.

(a) Staple yarn-many short fibers twisted together tightly (短纤纱—许多短纤维紧密地加捻在一起)

(b) Multifilament-many continuous filament with certain twist (复合长丝—许多连续长丝加上一定的捻度)

(c) Ply yarn-two single yarns twisted together (股线—两根单纱加捻在一起)

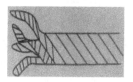

(d) Cord or cable-many plied yarns twisted into a coarse structure (粗线或缆绳—许多股纱线加捻成粗的结构)

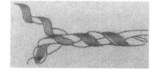

(e) Multiply yarn-two plied yarns twisted together (双股线—两根股线加捻在一起)

Fig 3.2 Structure of all kinds of yarns(各种纱线的结构)

【Text】

纱线可以按照多种方式分类。例如,根据是否在纺纱期间混合两种或更多种类型的纤维,纱线分为纯纺纱或混纺纱[1]。或者根据是否经过精梳工艺,纱线分为普梳纱或精梳纱[2],等等。

【Key Words】

pure yarn　纯纺纱
blended yarn　混纺纱
spinning ['spiniŋ]　纺纱

carded yarn　粗梳纱线
combed yarn　精梳纱线
combing process　精梳工艺

【Key Sentences】

1. Yarn can be classified into **pure yarn** or **blended yarn** depending on whether two or more types of fibers were mixed during **spinning.**

Fig 3.3 shows the pure yarn and blended yarn.

（a）Pure yarn（纯纺纱）　　　　　　（b）Blended yarn（混纺纱）

Fig 3.3　Pure yarn and blended yarn（纯纺纱与混纺纱）

2. Yarn can be classified into **carded yarns** and **combed yarns** depending on whether a **combing process** was involved.

【Text】

当纤维、短纤维或长丝形成纱线时,施加捻度来将纤维保持在一起[1]。加捻数量有时通过如低捻、中捻和高捻等术语广泛表示,更精确地用每英寸捻回数表示[2]。形成最佳纱线所需的每英寸捻回数(TPI)随纱线直径而变化[3]。纱线越细,需要的捻度越多。为了确定最佳捻度,采用参数——捻系数(TM)[4]。TM 的范围为 3(弱捻纱)~6 或更多(强捻纱)。可以使用捻系数来计算在任何支数中给出所需的纱线每英寸的捻回数(TPI),或者比较不同支数纱线的加捻程度[5]。纱线的强度部分归因于已施加的捻度,强捻纱线需要相当大的捻度。然而,超过最佳点,额外的扭曲会导致纱线扭结并最终失去强度[6]。

【Key Words】

construction [kənˈstrʌkʃn] 结构
medium [ˈmiːdiəm] 中等的,中级的
turns per inch 每英寸捻回数
optimum [ˈɒptiməm] 最适宜的
factor [ˈfæktə(r)] 因素
twist multiplier [ˈmʌltiplaiə(r)] 捻系数
soft yarn 弱捻纱
hard-twist yarn 强捻纱
calculate [ˈkælkjuleit] 计算,估计
count 数量
impart [imˈpaːt] 传授,告知
additional [əˈdiʃənl] 额外的,附加的
kink [kiŋk] 扭结

【Key Sentences】

1. As fibers, staple fibers or filament, are formed into yarns, twist is added to hold the fibers together.

2. The amount of twist is sometimes suggested broadly by such terms as low, **medium**, and high, but it is more accurately indicated by the number of **turns per inch**.

3. The turns per inch (TPI) needed to form the best possible yarn varies with the yarn diameter.

4. To determine the **optimum** degree of twist, a **factor** called the **twist multiplier** (TM) is used.

5. We can use the twist factor either to **calculate** the TPI required to give the desired degree of twist in any **counts**, or to compare the degree of twist in yarns of different counts.

6. Beyond an optimum point, **additional** twist will cause yarns to **kink** and finally to lose strength.

【Text】

在任何给定的短纤纱中赋予最大强度的捻系数有时被称为最佳捻系数[1]。捻向也很重要,纱线可以用左手捻(S 捻)或右手捻(Z 捻)进行加捻[2]。捻向符合字母的中心条。通过组合不同捻向的纱线可以获得各种效果,并且可以通过 S 捻和 Z 捻单纱的有效合股来提高耐久性[3]。

【Key Words】

maximum [ˈmæksiməm] 最大值的,最大量的
optimum twist factor 最佳捻系数
right-hand twist 右手捻(Z 捻)
direction [dəˈrekʃn] 方向
left-hand twist 左手捻(S 捻)

【Key Sentences】

1. The twist factor which gives the **maximum** strength in any given staple yarn is sometimes called the **optimum twist factor**.

2. The **direction** of twist is also important. Yarns can be twisted with either a **left hand twist** (S twist) or a **right-hand twist** (Z twist).

Fig 3.4 shows the yarn's **direction** of twist.

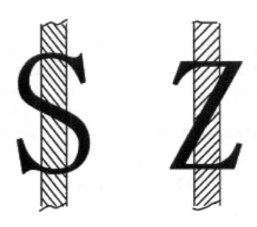

Fig 3.4　Yarn direction of twist(纱线捻向)

3. Various effects can be obtained by combining yarns of different twist direction, and durability may be increased by efficient plying of S and Z twist single yarns.

【Text】

纱线是按重量买卖的,但织造商对在给定重量下获得的纱线长度非常感兴趣[1];因为可以织制的织物量主要取决于可用于织造的纱线长度。因此,以每单位长度的重量或每单位重量的长度来描述纱线的细度或粗度[2]。以这些方式之一表示的纱线细度是纱线特数或支数[3]。在纺织工业已经建立的不同地区,随着时间的推移,已经发展了许多不同的纱线细度系统。它们都可以根据直接制(基于每单位长度的重量)或间接制(基于每单位重量的长度)进行分类[4]。

【Key Words】

vitally [ˈvaitəli] 极其,绝对　　　　　length per unit weight　单位重量长度
weight per unit length　单位长度重量　direct system　直接制
finenes [ˈfainnəs] 细度　　　　　　　indirect system　间接制
coarseness [kɔːsnəs] 粗度

【Key Sentences】

1. Yarns are bought and sold by weight, but the weaver is **vitally** interested in the length of yarns he will obtain in a given weight.

2. Therefore, it is natural to describe the fineness or coarseness of yarns in terms of **weight per unit length**, or of **length per unit weight**.

3. The fineness of a yarn expressed in one of these ways is its yarn number or **count**.

4. They all can be classified under the headings of **direct** (based on weight per unit length) or **indirect systems** (based on length per unit weight).

Project Three　Yarns and Spinning Technology(纱线与纺纱技术)

【Text】

常用的直接制有旦尼尔制和特克斯制。旦尼尔制是表示纤维、长丝或纱线9000m长时的重量克数,或者9m长时的毫克数[1]。特克斯制是一种建议的通用系统,旨在取代所有其他系统,纱线的特克斯制是表示1000米长纱线的重量克数[2]。

显然,旦尼尔和特克斯之间存在简单的关系,即1旦尼尔 = 9特克斯。在任何直接制中,单位长度的重量与数值成正比,较粗的纱线细度意味着数值越高[3]。在直接制中,纱线细度是可以相加的,例如,复合纱线的细度是各组纱线细度的总和[4]。直接制的数值越大,纱线越粗[5]。

【Key Words】

denier ['deniə(r)] system　旦尼尔制
tex system　特克斯制
universal [ˌjuːniˈvɜːsl] 通用的
eventually [iˈventʃuəli] 终究,终于,最后
namely [ˈneimli] 即,也就是
equivalent [iˈkwivələnt]
alternative [ɔːlˈtɜːnətiv] 替代的,备选的
state [steit] 规定,陈述

particular [pəˈtikjələ(r)] 特别的,独有的
proportional [prəˈpɔːʃənl] 比例的,成比例的
additive [ˈæditiv] [数] 加法的
composite [ˈkɔmpəzit] 复合的
component [kəmˈpəunənt] 成分,组分
two-fold [ˈtuːfəuld] 双重的,两倍的
take-up　卷绕

【Key Sentences】

1. The **denier system**: The denier of a fiber, filament or yarn is defined as the weight in gram of 9,000 m or, alternatively, the weight in microgram of 9 m.

2. The **tex system**: This is the proposed **universal** system which is intended **eventually** to replace all other systems. The tex number of a yarn is the weight in gram of 1,000 m.

3. In any direct system, the weight per unit length is directly **proportional** to the yarn number, and a higher yarn number implies a coarser yarn.

4. Yarn numbers expressed in a direct system are **additive**, i.e., the number of a **composite** yarn is the **sum** of the numbers of its **component** yarns.

5. The larger the value of the direct system, the thicker the yarn.

【Text】

常用的间接制有公制支数和英制支数。公制支数是指1克重的纱线在公定回潮率下的长度米数。棉英制即棉纱支数可定义为绞数(每绞840码),即1磅重绞数(840码的倍数),例如,30支棉纱有30×840 = 25200码/磅[1]。精纺毛英制即精纺毛支数定义为绞数(每绞560码),即1磅重绞数(560码的倍数),例如,30支精纺毛纱为30×560 = 16800码/磅[2]。

在任何间接制中,纱线支数与每单位长度的重量成间接(反)比例,支数越高意味着纱线越细[3]。间接纱线支数量不可以直接相加,为了获得复合纱线的最终支数,必须使组分纱支数的倒数相加并取其总和的倒数[4]。例如,由20支的一根纱线和30支的一根纱线组成的双股棉纱的

最终支数为12支,因为1/20+1/30=1/12。

【Key Words】

indirect system　间接制(定重制)
cotton system　棉英制
hank [hæŋk]　绞
yd 码(1 码=0.9144 米)
1b 质量单位:磅

worsted [ˈwəːstid] system　精纺毛英制
indirectly proportional　反比例
additive　相加的
resultant　因而发生的,结果必然产生的
reciprocal [riˈsiprəkl]　倒数

【Key Sentences】

1. The **cotton system**: cotton count is defined as the number of **hanks**, each 840 **yd.** long, which weigh 1 **1b**. For example, a 30^s cotton count yarn has 30×840 = 25,200 yd./lb.

2. The **worsted system**: worsted count is defined as the number of hanks, each 560 yd. long, which weigh 1 lb. For example, a 30^s worsted count yarn has 30×560 = 16,800 yd./1b.

3. In any indirect system, the yarn number is **indirectly proportional** to the weight per unit length of the yarn, and a higher number implies a finer yarn.

4. Indirect yarn numbers are not **additive**. To find the **resultant** count of a folded yarn, one must add the **reciprocals** of the component counts and take the reciprocal of the sum.

Part 2　*Trying*

1. Sort the pure cotton yarn of the following four different tex numbers from thin to coarse.

Sample	20 tex	16 tex	36 tex	48 tex
Sort				

2. Sort the pure cotton yarn of the following four different Nm numbers from coarse to thin.

Sample	30 Nm	20 Nm	48 Nm	18 Nm
Sort				

3. Sort the pure cotton yarn of the following four mixed fineness units from coarse to fine.

Sample	20 tex	30 Nm	36 tex	20 Nm
Sort				

Part 3　*Thinking*

1. In the various types of garment fabrics seen in daily life, which garment fabrics use blended yarn? Why they use blended yarn?

2. In terms of the clothing (fabrics) you wear by yourself, which fabrics require coarse yarns and which fabrics require fine yarns?

Task Two Opening and Cleaning Technology(开清棉技术)

【Text】

短纤维以高度压缩的棉包(或化纤包)到达纱线加工厂[1]。在纺成纱线之前,必须将大块棉花彻底松开并打开[2]。为了纺制纱线,纤维必须具有相似的长度和相对均匀的性能,以使细纱具有均匀的质量[3]。为了完成这一点,不同生产批号、来自不同产地和动物身上的纤维必须先混合在一起。因此,纺纱的第一道工序称作开清棉。在开清棉车间,重要的是要分散并"开松"棉(纤维)团至单根纤维状态,或则尽可能地接近这种单纤维状态[4]。开棉机和混棉机将从不同棉包(箱)取来的棉纤维开松并混合[5]。抓棉机从棉包或化纤包中抓取棉束和棉块喂给开棉机和混棉机,但抓取的同时也有开松、混和作用[6]。

【Key Words】

bale [beil] 大包
compress [kəmˈpres] 压缩,压紧
lump [lʌmp] 块,团
throughly [ˈθruːli] 十分地,彻底地
loosen [ˈluːsn] 放宽,放松
lots [lɔts] 此处译为:批号、标签
meet 满足

carton [ˈkaːtn] 棉箱
opening and cleaning 开清棉
opening machine 开棉机
blending machine 混棉机
workshop 车间
(bale) plucker 抓棉机

【Key Sentences】

1. Staple fibers arrive at the yarn processing plant in large **bales** that are so highly **compressed**. Fig 3.5 shows the cotton bale.

2. Large **lumps** of cotton must be **throughly loosened** and **opened out** before spinning to yarn.

3. To make yarns, fibers must be of similar length and relatively uniform, so that the spun yarn can be of uniform quality.

4. It is important to separate or "open" the fiber mass to a single fiber state, or as close to that

Fig 3.5　Cotton bale(棉包)

as possible in opening and cleaning workshop.

5. The **opening** and **blending machines** separate the fibers and blend fibers from the different bales or **cartons**.

6. The **plucker** grabs the bundle and the lump from the cotton bale or the chemical fiber bale and feeds them to the opener and the cotton blender, but it also has the function of opening and mixing while grabing.

Fig 3.6 shows different bale pluckers.

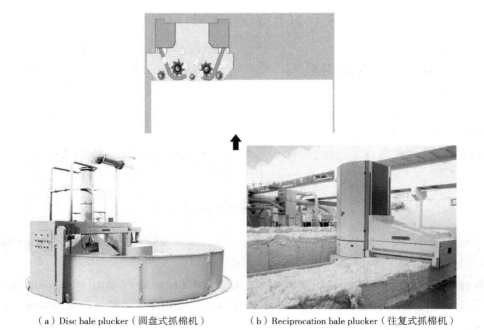

（a）Disc bale plucker（圆盘式抓棉机）　　　（b）Reciprocation bale plucker（往复式抓棉机）

Fig 3.6　Bale plucker(抓棉机)

Project Three Yarns and Spinning Technology(纱线与纺纱技术)

【Text】

开棉是将大块的棉团分解为小的纤维块[1]。当棉包喂入纺纱厂的第一道加工设备,棉花处于坚硬的、压紧的簇状或纠缠的状态[2]。为了使棉花呈现良好的状态进入后道设备,以保证纺出清洁、均匀的纱线,必须将成簇的棉团进行开松[3]。所以,开棉机械的主要作用是对原料进行开松和除杂。原棉的开松和除杂作用过程是相辅相成的:在将原棉松解成小棉束的同时,使纤维与杂质分离,通过机械运动实现少量的除杂作用[4]。这些任务是通过不同种类的开棉机完成的。

清棉是通过机械方式除去无用的杂质[5]。这是将有用的棉纤维和杂质分离开的最有效和经济的方式,且对纤维没有过多的损伤[6]。虽然清棉和开棉在功能上的定义有些不同,实际上,它们是在同一台机器上进行的,也正是因为这个原因,开、清棉被认为是一体的[7]。目前,许多纺纱企业使用单独的清棉机对原棉进行处理。

【Key Words】

break up 使散开
tuft [tʌft] 块状,簇状
mill [mil] 工厂
mattted ['mætid] 缠结的,纠缠的
tangled ['tæŋgld] 缠结的,混乱的
subsequent ['sʌbsikwənt] 随后的,就下来的
blooming ['bluːmiŋ] 盛开的,开松的
complementary [ˌkɔmpli'mentri] 相辅相成的
removal [ri'muːvl] 除去

trash [træʃ] 杂质
efficiently [i'fiʃntli] 有效地,效率高地
economically [ˌiːkə'nɔmikli] 经济地,节约地
usable ['juːzəbl] 可用的
lint [lint] 棉纤维
undue [ˌʌn'djuː] 过度的,过分的
currently ['kʌrəntli] 通常,一般
mechanism ['mekənizəm] (机械)结构,机械装置

【Key Sentences】

1. Opening is the **breaking up** of the fiber mass into **tufts**.

Fig 3.7 shows three different cotton opening machines.

2. When the bales of cotton are brought to the first processing machine in the spinning **mill**, cottons are in hard, compressed, **mattted** and **tangled** form.

3. To make sure the cottons fed to **subsequent** machines in good condition to produce clean and uniform yarn, there must be an opening up of **blooming** matted cottons.

4. The process of opening and removing impurities of raw cotton is **complementary**. When the raw cotton is loosened into small cotton bundles, the fibers are separated from the impurities, and a small amount of impurity removal is achieved by mechanical movement.

5. Cleaning is the **removal** of useless **trash** by mechanical means.

Fig 3.8 shows the cleaning machine.

6. It is the process of **efficiently** and **economically** separating the **usable lint** from the trash, and

 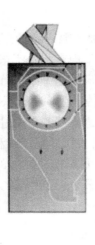　　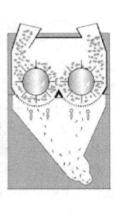

（a）Single axis flow opener（单辊筒轴流式开棉机）　　　　（b）Double axis flow opener（双辊筒轴流式开棉机）

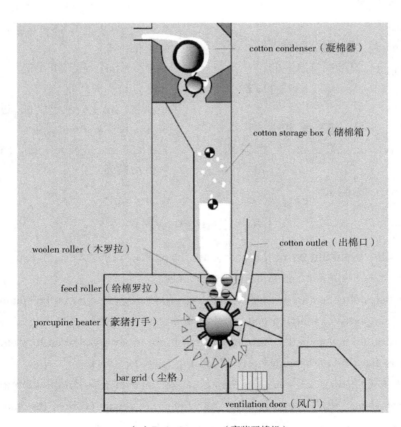

（c）Porcupine opener（豪猪开棉机）

Fig 3.7　Three different cotton opening machines（三种不同的开棉机）

Project Three　Yarns and Spinning Technology(纱线与纺纱技术)

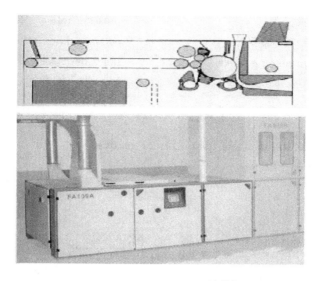

Fig 3.8　Cleaning machine(清棉机)

it contains no **undue** damage to fibers.

7. While cleaning defined here as a function different from the opening, actually, the opening and cleaning are accomplished **currently** by the same **mechanism** and, so as to be considered as a whole.

【Text】

混棉是将不同等级的纤维混合[1]。来自不同棉包的纤维是有所区别的,为了使所纺纱线的质量比较均匀,通常要将不同棉包的纤维进行混合[2]。混棉作用也是通过专门的混棉机械完成的,常用的混棉机械有棉箱式混棉机和多仓式混棉机。棉箱式混棉机一般有较大容积的储棉箱,机内装有角钉帘,对抓棉机喂入的原料进行扯松与混和[3];多仓混棉机一般有6~8个储棉仓,利用"时间差""行程差"对纤维进行充分混和[4]。

混合后的纤维在天平调节装置下被气流均匀送入单打手成卷机,棉束接受高速回转的打手的撕扯和打击,进一步去除杂质[5];而开松后的棉纤维吸附在回转的双尘笼表面,经尘笼滚压形成棉层,此时细小尘杂和短绒透过尘笼表面网眼由两侧风道排出[6]。棉层经剥棉罗拉、棉卷凹凸防粘罗拉、紧压罗拉、导棉罗拉的引导作用并最终输送给棉卷罗拉,卷绕在棉卷辊上,制成一定长度、均匀度的棉卷[7]。

【Key Words】

blending　混棉
mix　混合
grade　等级
cylinder ['silində(r)]　锡林
roll　压,辗
hopper blending machine　棉箱混棉机
corner nail lattice　角钉帘
multi-bin mixer/blender　多仓混棉机
time difference　时间差
path difference　路程差
single beater scutcher　单打手成卷机
balance adjusting device　天平调节装置

doubledrum cage 双尘笼
cotton stripped roller 剥棉罗拉
split-lap preventer en bossing roller 棉卷凹凸防粘罗拉

pressing roller 紧压罗拉
cotton guiding roller 导棉罗拉
lap roller 棉卷罗拉
lap roll 棉卷辊

【Key Sentences】

1. Blending is the **mixing** of fibers from different **grades**. Fig 3.9 shows different blenders.

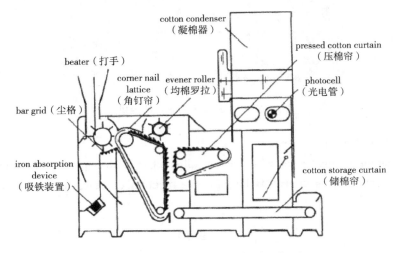

(a) Hopper blending machine（棉箱式混棉机）

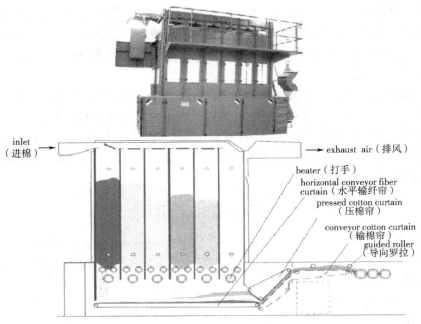

(b) Multi-bin mixer/blender（多仓混棉机）

Fig 3.9　Different blenders（不同混棉机）

2. Cotton varies from bale to bale, so the fibers from several bales are blended together to give yarn of more uniform quality.

3. **The hopper blending machine** generally has a large volume cotton storage box, and the machine is equipped with **corner nail lattice** which loosen and mix the raw materials fed by the bale plucker.

4. **Multi-bin blender** generally has 6 to 8 cotton storage bins, which use the "**time difference**" and "**path difference**" to fully mix the fibers.

5. The mixed fiber is uniformly fed into the **single beater scutcher** by the airflow through the **balance adjusting device**, and the cotton bundle receives the tearing and striking of the high-speed rotating beater to further remove the impurities.

Fig 3.10 shows process schematic of single beater scutcher.

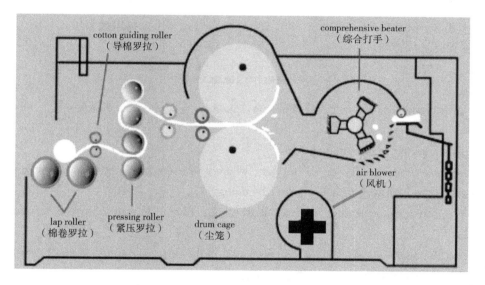

Fig 3.10 Process schematic of single beater scutcher(单打手成卷机的加工原理图)

6. the opened cotton fibers are adsorbed on the surface of the rotating **double drum cage** and are rolled to form the cotton layer. while fine dust and short fibers passing through the surface of the double drum cage and then discharged from the air passages on both sides.

7. The cotton layer is guided by the **cotton stripped roller**, **split-lap preventer en bossing roller**, the **pressing roller** and the **cotton guiding roller**, and finally conveyed to the **lap roller** and wind on the **lap roll** which make the picker lap with a certain length and uniformity.

【Text】

在纺纱系统中,上面所讲的几个设备的连接称作间歇式纺纱系统。在间歇性纺纱系统中,开松机器将纤维分离成蓬松的纤维团[1];这些松散的纤维被送入喂料斗,料斗称得一定量后放置在传送带并送至混棉装置,进一步进行混合[2];清棉装置进一步通过罗拉和高压气流系统对纤维进行开松、除杂和混合[3];最后,棉纤维经过成卷机加工成棉卷,然后将棉卷运送到梳棉装置进行下一步处理,可见开清棉工序与梳棉工序是完全独立分开的,没有直接连接在一起。

在开清棉工序中,大部分尘土和杂质通过重力和离心力的作用被除去[4]。棉纤维将比再生纤维素纤维经历更多开松和除杂程序,因为棉纤维与再生纤维素纤维相比,杂质更多,不均匀度更大[5]。

【Key Words】

gravity ['grævəti] 重力,万有引力,地心引力
centrifugal [ˌsentri'fju:gl] force 离心力
variation [ˌveəri'eiʃn] 变化,变动
intermittent [ˌintə'mitənt] system 间歇性纺纱系统
fluffy ['flʌfi] 松软的
hopper ['hɔpə(r)] 料斗

conveyor [kən'veiə(r)] belt 传送带
blending unit 混棉装置
picker 清棉机
roller 罗拉
forced air 高压气流
picker lap 棉卷
card unit 梳理装置

【Key Sentences】

1. In the **intermittent system**, the opening operation separates the fibers into a loose, **fluffy** mass.

Fig 3.11 shows the intermittent system.

2. These loose fibers are fed into a **hopper**, where a measured amount is laid on a conveyor belt and delivered to the **blending unit** for further blending.

3. The picker further opens, cleans, and blends fibers through a system of **roller**s and **forced air**.

4. During the opening and cleaning process, most of the dirt and impurities that might be present are removed by either **gravity** or **centrifugal force**.

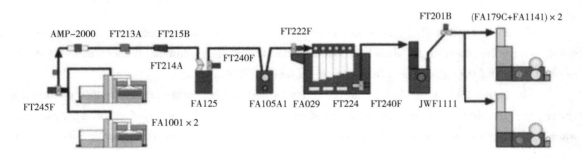

Fig 3.11　Intermittent system[间歇式系统(纤维→棉卷)]

5. Cotton fibers receive more opening and blending than man-made fibers, since they have more impurities and greater **variation** than do man-made fibers.

Project Three　Yarns and Spinning Technology(纱线与纺纱技术)

【Text】

在生产环锭纱线时,如果采用连续加工系统,则纤维从棉包开始,要自动加工至少到梳棉棉条(生条)阶段[1]。这就说明,连续加工系统取消了成卷工序,也将连续加工系统称作清梳联系统。在自动装置(连续加工系统)上生产的纱线与用间歇性系统生产的纱线相比,条干更均匀,强度更高[2]。连续化生产的速度较高,节省人力,工厂也比较清洁[3]。不管采用怎样的加工系统,纱线的最终质量主要取决于原料的选择和开棉、除杂、混棉的彻底性以及清棉加工[4]。在间歇性系统中,清棉棉卷放置在梳棉机的机尾处,为梳棉机提供原料[5];如果采用自动化的加工系统,纤维则通过喂棉箱以松散纤维的形式直接喂入梳棉机[6]。

【Key Words】

continuous [kən'tinjuəs] system　连续生产系统
blowing-carding system　清梳联系统
ring-spun yarn　环锭纺纱线
automatically [ˌɔːtə'mætikli]自动地
card sliver　生条
state　状态
move on　前行,继续前进(此处可翻译为继续加工)
discontinuous [ˌdiskən'tinjuəs]　非连续的,不连续的
thoroughness ['θʌrənəs]完全,十分
rear　后部,背面,背后
automatic system　自动系统

【Key Sentences】

1. The **continuous system** for producing **ring-spun yarns** takes fibers directly from the bale and processes them **automatically** through to the **card sliver state** at least.

Fig 3.12 shows the continuous system(blowing-carding system).

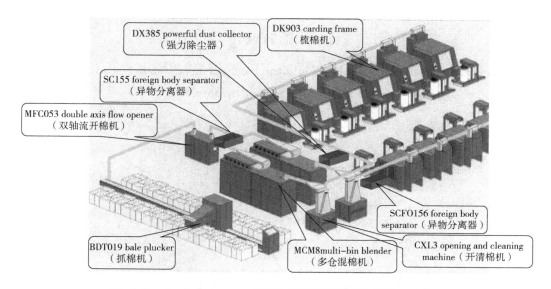

Fig 3.12　Continuous system(连续式系统即清梳联:纤维→生条)

2. Yarns made on automatic equipment tend to be more uniform and may be stronger than **discontinuous** process yarn.

3. Production speed is considerably faster for continuous processes, labor costs are reduced, and plants stay cleaner.

4. No matter which system is used, the quality of the final yarn depends largely on the selection of fibers and on the **thoroughness** of the opening, cleaning, blending, and picking operations.

5. In the intermittent or discontinuous systems, the **picker lap** is placed at the **rear** of the card frame to supply fibers.

6. In the **automatic system**, the fibers are held in a hopper and fed in a loose form directly to the carding machine.

Fig 3.13 shows the picker lap.

Fig 3.13　Picker lap[棉卷(梳棉机)]

Part 2 *Trying*

1. Please represent the difference between intermittent spinning system and continuous spinning system in the process of opening and cleaning technology by drawing.

2. Please represent the blending principle of the following blending machine by simple illustration.

Project Three Yarns and Spinning Technology(纱线与纺纱技术)

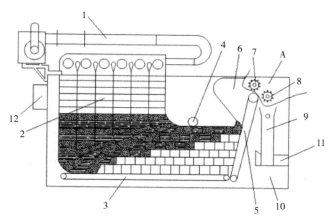

1—cotton feeding pipe(喂棉管道) 2—cotton bin(棉仓) 3—horizontal conver belt(水平导带) 4—monitoring device(监控装置) 5—corner nail lattice(角钉帘) 6—small blending box(小混棉箱) 7—evener roller(均棉罗拉) 8—stripping roller(剥棉罗拉) 9—cotton storage tube(储棉管) 10—dropping impurity storage tan(落棉杂质储存箱) 11—cotton storage table(储棉台) 12—air recovery outlet(气流回收出口)

Part 3 *Thinking*

1. Summarize what is the main role of the opening and cleaning technology?

_____
_____
_____

2. According to the learned knowledge of textile materials, which fibers do you think are more demanding for the opening and cleaning technology during spinning?

_____
_____
_____
_____

Task Three   Carding and Combing Technology (梳棉与精梳技术)

【Text】

连续纺纱系统中,开清棉工序以后紧接着的是梳棉工序。在大多数纺纱工艺中,它都是一个很重要的工序,在棉纺、棉废纺、精梳毛纺、半精梳毛纺、粗梳毛、黄麻和亚麻纺纱系统当中,不论是天然纤维纺纱还是再生纤维纺纱都会用到梳棉机[2]。梳棉是将纠缠的纤维块分梳成单纤维组成的薄纤维网,在一个密封的空间中,靠包覆有反向锋利针齿的针布表面来完成梳理纤维块的动作[3]。

【Key Words】

tow system [təuˈsistəm] 短麻纺
employ [imˈplɔi] 使用,利用
carding machine 梳棉机
filmy [ˈfilmi] 薄的
opposing 相反的
sharp points 锋利的针齿(尖)

【Key Sentences】

1. In the continuous spinning system, opening and cleaning operations are followed by **carding**.

2. It is an important operation in most spinning processes: cotton, cotton waste, worsted, semi-worsted, woolen, jute and flax **tow systems** all **equipped with carding machines** either for natural fibers spinning or man-made fibers spinning.

Fig 3.14 shows the carding machine.

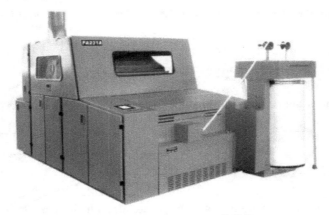

Fig 3.14　Carding machine(梳棉机)

3. Carding may be defined as the process to reduce tufts of entangled fibers into a **filmy** web of individual fibers, the action of which is performed by the surface of card clothing covered with **opposing sharp points**.

Fig 3.15 shows the sharp points.

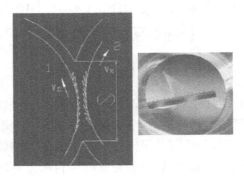

Fig 3.15　Sharp points(锋利的针齿)

【Text】

梳棉工序的任务如下:将纤维块分解为单纤维;除去杂质和短绒;充分混合纤维;形成生条[1]。梳棉是靠针布完成的,针布指的是覆盖在罗拉表面的大量的梳针[2]。梳针主要有两种:安装在底布上的弹性针和刚性金属针[3]。刚性金属针在梳棉机上应用广泛。两组针以相同的方向、不同的速度运动,将纤维梳理成薄层状,在锡林上形成薄薄的纤维网[4]。这层薄网经集束后,拉伸成须条,叫作生条[5]。生条经牵引通过一个漏斗,然后圈落到条筒里或者送到传送带上[6]。

【Key Words】

summarize [ˈsʌməraiz] 总结,概述
sliver [ˈslivə(r)] 条子
card clothing 针布
pin 针
mount into 安装,架构
foundation [faunˈdeiʃn] 地基(此处译为基布)

rigid metallic card cloting 刚性金属针布
tease [tiːz] 梳理
ropelike 像绳子一样
funnel [ˈfʌnl] 漏斗
doff 络纱
conveyor belt 运输带,传送带

【Key Sentences】

1. The tasks of carding may be **summarized** as follows: separating fiber bundles into individual, removing dusts and lints, intimately blending fibers, formation of **sliver**.

2. Carding is accomplished by **card clothing** which is referred as the large number of **pins** covering the roller surface.

3. There are two main types of card clothing: flexible wire **mounted into** a **foundation** and **rigid metallic** card clothing .

Fig 3.16 shows rigid metallic card clothing and flexible wire.

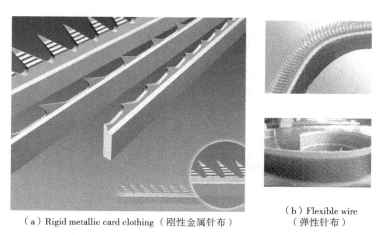

(a) Rigid metallic card clothing (刚性金属针布)　　(b) Flexible wire (弹性针布)

Fig 3.16　Rigid metallic card clothing and flexible wire(刚性金属针布与弹性针布)

4. The two sets of pins move in the same direction, but at different speeds, to **tease** the fibers into a filmy layer; so that a thin web of fibers is formed on the cylinder.

5. This thin web is gathered into a soft mass and pulled into a **ropelike** strand of fibers, which is called sliver.

6. The sliver is pulled through a **funnel** and then **doffed** or delivered to cans or to a **conveyor belt**.

【Text】

梳棉机是由锡林、盖板、刺辊和道夫组成的,道夫表面由植入特制弯针的织物所覆盖[1]。梳棉机的主要组成部分是一个滚筒状的锡林和带有窄的金属条块的盖板。梳棉机的功能是改善纤维的排列取向,除去残留的杂质,同时除去一些较短的和不成熟的纤维,这主要是由锡林和盖板完成的[2]。

【Key Words】

cylinder ['silində(r)] 锡林
flat 盖板
taker-in 刺辊
doffer [dɔfə] 道夫

embed [im'bed] 把…嵌入
drum [drʌm] 鼓
strap [stræp] 带子
arrangement [ə'reindʒmənt] 安排,排列

【Key Sentences】

1. The carding machine consists of **cylinders**, **flats**, **taker-in** and **doffers** covered with heavy fabric **embedded** with specially bent wires.

Fig 3.17 shows the sectional structure of the conventional carding machine.

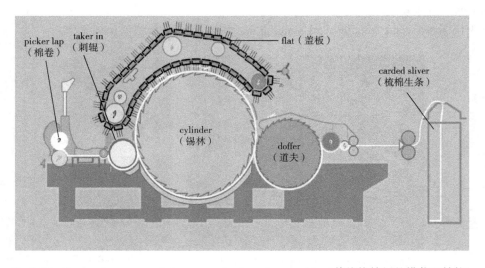

Fig 3.17 Sectional structure of the conventional carding machine(传统梳棉机的横截面结构)

2. The function of the card is to improve the **arrangement** of the fibers, remove all the remaining impurities and at the same time very short and immature fibers. This is done primarily by the cylinder and flats.

【Text】

梳棉机主要包含三个部分:给棉与刺辊部分、锡林和盖板部分及凝聚和输送部分[1]。给棉与刺辊部分可以被看作是锡林和盖板区的一个准备部分。在棉纺工序中,大量的纤维在刺辊部分分离成单纤维,很多杂质也由除尘刀和刺辊下方的小露底清除掉[2]。锡林和盖板部分也叫作梳理部分,这是梳棉工序的主要部分,锡林和盖板的作用是:使原料完全开松,甚至分离成单纤维状态;收集短纤维和杂质,使它们和长纤维分离开来[3]。凝聚和输送部分的目的是:由道夫从锡林上收集纤维;由道夫斩刀或其他装置从道夫上面将纤维剥离;借助紧压罗拉和喇叭口将纤维集聚为连续的须条,就是大家熟知的生条;用圈条器将生条收集在条筒中[4]。要想制成高质量的生条,这三个运动部分是非常重要的,因为它们是相互关联的。

【Key Words】

feeding and taker-in part 给棉与刺辊部分
cylinder and flats part 锡林和盖板部分
condensing and delivering part 凝聚和输送部分
mote [məut] knife 除尘刀
casing ['keisiŋ] 壳,套,罩(此处译为小露底)
object ['ɔbdʒikt] 目标
strip off 剥棉
condense [kən'dens] 凝聚
calendar roller 紧压罗拉
trumpet ['trʌmpɪt] 喇叭口
deposit [di'pɔzit] 寄存,放置,安置
coils [kɔilz] 圈条

【Key Sentences】

1. In carding machine, there are chiefly three parts: **feeding and taker-in part**, **cylinder and flats part**, and **condensing and delivering part**.

2. A great part of fibers such as in cotton spinning are separated into single fibers by the taker-in part, and also a great part of trash and impurities are removed by **mote knife** and **casing** in the taker-in region.

Fig 3.18 shows the feeding and taker-in part.

3. The purposes of the cylinder and flats are: to open the raw materials completely, even to separate one fiber from all the others; to collect short fibers and dirt, and to separate them from longer fibers.

Fig 3.19 shows the cylinder and flats part.

4. The objects of condensing and delivering part are: to collect fibers from the cylinder by a smaller cylinder known as the doffer, to **strip off** fibers from the doffer by doffer comb or any other devices, to condense them into a continuous rope-like form known as sliver with the help of **calendar**

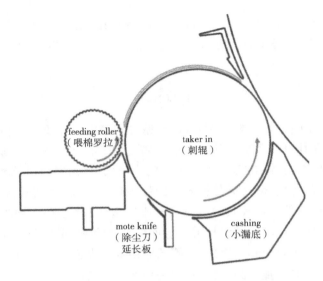

Fig 3.18　Feeding and taker-in part(给棉刺辊部分)

Fig 3.19　Cylinder and flats part(锡林盖板部分)

rollers and **trumpets**, to **deposit** slivers in the form of coils in a can with the help of coils mechanism.

Fig 3.20 shows the condensing and delivering part.

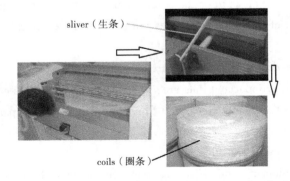

Fig 3.20　Condensing and delivering part(凝聚和输送部分)

Project Three Yarns and Spinning Technology(纱线与纺纱技术)

【Text】

一些高档棉织物,需要精梳纱而不是粗梳纱(普梳纱)[1]。例如,当棉与再生纤维混纺时,通常二者混合之前,棉纤维要经过精梳工序。梳棉生条中含有一些小颗粒(杂质),少量的棉结和限定为长度在 16 mm 以下的棉短纤或 30mm 以下的羊毛短纤[2]。此外,一些梳棉生条中的单根纤维没有很好的纵向排列,大部分纤维在一端勾结,也有许多纤维在两端勾结[3]。用于去除短纤及残留杂质的工序称作精梳。精梳的基本目的是:去除短纤维;去除如棉结或粗节等疵点的非纤维状杂质;伸直与平行保留的长纤维。精梳是一种达到或增强理想的纱线性能的手段。精梳可使最终纱线比其他方式所得纱线更光滑、更精细、更坚固、更有光泽、更均匀[4]。

【Key Words】

finer quality   优质
combed yarn   精梳纱
nep   棉结
remnant [ˈremnənt]   残留的
impurity [imˈpjuərəti]   杂质
align [əˈlain]   使成一条直线

longitudinally [ˌlɔndʒiˈtjuːdinli]   纵向,纵向地
hook [huk]   勾在一起
non-fibrous   非纤维状的
straighten [ˈstreitn]   使伸直
parallelize   使平行

【Key Sentences】

1. Fine quality cotton fabrics require **combed yarns** rather than carded yarns.

2. Card sliver contains some small particles, small quantities of **neps**, and short fibers which are defined as being less than 16 mm in length for cotton, and 30 mm for wool.

3. In addition, single fibers in card sliver are not well **aligned longitudinally** and most of them are **hooked** on one end and many are hooked on both ends.

4. Combing enables the ultimate yarn to be smoother, finer, stronger, more lustrous, and more uniform than otherwise would be possible.

Fig 3.21 shows the comparison of surface morphology between combed yarn and carded yarn.

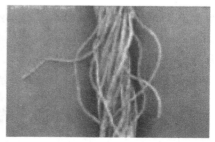

(a) Combed yarn (精梳纱)   (b) Carded yarn (普梳纱)

Fig 3.21 Comparison of surface morphology between combed yarn and carded yarn
(精梳纱与普梳纱的表面形态对比)

【Text】

　　一般来说,梳理仅限于长度较长和纤维长度更均匀的较好等级的天然纤维[1]。它用于生产精梳棉纱、精纺毛纱、一些亚麻纱线和绢纺纱。40 支甚至更细的细棉纱,通常由精梳棉纺制而成,满足质量均匀和外观均匀的最终用途[2]。当强度至关重要时,即使是粗纱也需要精梳棉。许多针织产品都是用精梳棉制成的,以获得最佳的外观和手感。

　　根据纱线的最终用途,由精梳机除去的落棉(废物)量可以在 5%~25% 范围内[3]。除去高达 9% 的落棉,纱线可称为半精梳。对于大部分精梳纱线,去除的落棉率在 10%~18% 范围内,而在全精梳的情况下,其可能高达 25%[4]。这通常意味着用双精梳来获得最高质量的纱线——在第一次精梳中去除 18%,在第二次精梳中去除 7%[5]。因此,精梳工序和长纤维的成本都很高;由精梳机下来的须条叫作精梳棉条[6]。

【Key Words】

as a general rule　　一般来讲
be confined to　　局限于,限制于
worsted yarn　　精纺毛纱
spun silk yarn　　绢纺纱
noil　　精梳短毛,落棉,落麻
half combed　　半精梳
full combing　　全精梳
double combing　　全精梳
costly　　成本高的
emerge [iˈməːdʒ]　　显露出,呈现
combed sliver　　精梳条

【Key Sentences】

1. **As a general rule**, combing **is confined to** the better grades of natural fibers which have a longer and more uniform fiber length.

2. Fine cotton yarns, 40ˢ and finer, are usually spun from combed cotton when uniformity of mass and appearance are required for the end use.

3. The amount of **noil** (waste) removed by the combing machine can be in the range of 5 to 25 percent depending on the end use of the yarn.

4. For the largest proportion of combed yarn, the noil removed is in the range of 10 to 18 percent, whereas in the case of **full combing**, it may as high as 25%.

5. This often means **double combing** is to obtain the highest quality yarns——18% removed in the first combing and 7% removed in the second.

6. Fiber strand **emerging** from the combing machine called **combed sliver**.

【Text】

　　从梳棉机下来的梳棉生条,其形式与纤维排列都不适合(直接)精梳[1]。因此,在精梳前,必须做一些准备工序。在棉精梳准备工序中,梳棉生条经牵伸并形成小卷,然后再精梳[2]。精梳准备工序的目的是:改善纤维在须条中的平行度、伸直度与分离度;减少精梳加工中好纤维的损失;减少精梳加工中纤维的损伤;将生条变成合适的形式(提供给精梳机)。目前已有几种不同

的两步小卷制备方法。三种主要方法包括:并条机后面是条卷机;条卷机后面是并卷机;并条机后面是条并卷联合机[3]。这三种方法对应的分别是条卷工序、并卷工序和条并卷工序。但无论是哪种精梳准备工序,其最终目的都是将生条加工成符合要求的小卷[4]。

【Key Words】

as regards　至于,提到
preparatory [priˈpærətri]　预备的,筹备的
parallelization　平行度
straightness　伸直度
separation　分离度
lap　译为小卷
sliver lap machine　条卷机
ribbon lap machine　并卷机
draw frame and lap machine combined　条并卷联合机
sliver-lap process　并条—条卷工序
lap-lap process　条卷—并卷工序
sliver-lap-lap process　并条—条并卷工序

【Key Sentences】

1. The card sliver down from the card is unsuitable for combing both **as regards** form and fiber arrangement.

2. In cotton combing **preparatory** process, the card sliver is drawn and formed into a lap, which is then combed.

3. There have been several different two-step lap preparation methods. The three main methods include: the drawframe followed by a **sliver lap machine**; the sliver lap machine followed by the **ribbon lap machine**; the drawframe followed by **draw frame and lap machine combined**.

Fig 3.22 shows the sliver lap machine, Fig 3.23 shows the sliver lap machine, Fig 3.24 shows the draw frame and lap machine combined.

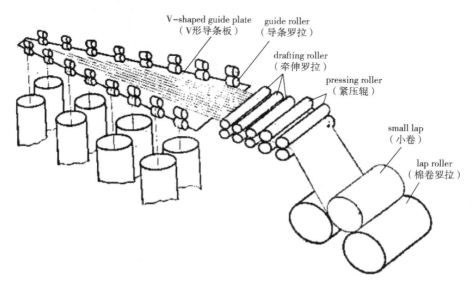

Fig 3.22　The sliver lap machine(条卷机)

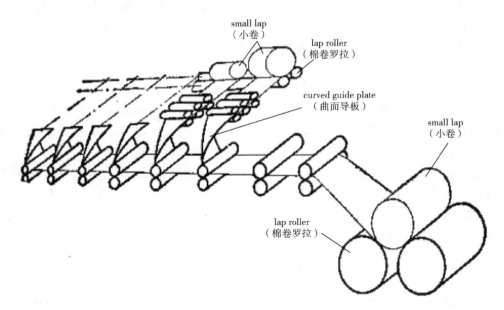

Fig 3.23 The ribbon lap machine(并卷机)

4. But no matter which the combing preparation process is, the ultimate object is to process the sliver into a lap that meets the requirements.

Fig 3.24 shows three kinds of combing preparation process.

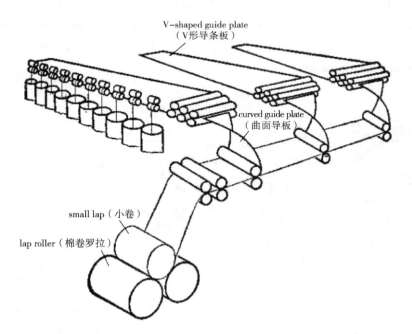

Fig 3.24 The draw frame and lap machine combined(条并卷联合机)

Project Three Yarns and Spinning Technology(纱线与纺纱技术)

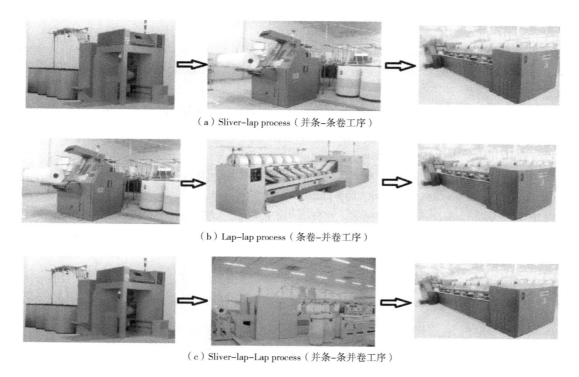

(a) Sliver-lap process (并条-条卷工序)

(b) Lap-lap process (条卷-并卷工序)

(c) Sliver-lap-Lap process (并条-条并卷工序)

Fig 3.25 Combing preparation process(精梳准备工序)

【Text】

精梳工序,在广义上,包括将棉条卷成连续的须条片状物,将须条的两端用金属梳子梳理,并将梳理的须条重新组合成新的棉条[1]。精梳机一般将6~8个小卷同时喂入,经梳理后棉条合并后经终牵伸成精梳条。精梳机一般间歇运行,因为长度短的纤维不允许使用连续的精梳方法[2]。须条的一端用圆形的锡林梳理,而另一端用单独排列的针(顶梳)进行梳理。须条的两端各自被梳理后,通过接头系统将分离的须条重新结合。精梳机的内部结构组成部分主要包括给棉罗拉、上钳板、下钳板、精梳锡林、顶梳、上分离罗拉、下分离罗拉等。精梳锡林与顶梳是梳理的主要部件。

【Key Words】

in broad terms of　从广义上讲
tuft [tʌft]　簇状物
reassemble [ˌriːəˈsembl]　重新组合
intermittently　间歇地
circular [ˈsəːkjələ(r)]　圆形的
top comb　顶梳
re-unit [ˌriːjuːˈnait]　再结合

piece [piːs]　结合,接头
internal [inˈtəːnl]　内部的
feed roller　给棉罗拉
lower nipper　下钳板
upper nipper　上钳板
top detaching roller　上分离罗拉
bottom detaching roller　下分离罗拉

【Key Sentences】

1. The combing operation consists **in broad terms of** taking the sliver to pieces in successive **tufts**, combing both ends of the tuft with metal combs, and **reassembling** the combed tufts into a new sliver.

2. The combed frame operates **intermittently** because the short length of the fiber does not allow the use of continuous combing.

Fig 3.26 shows the combed frame.

3. One end of the cotton beard is combed by a **circular** cylinder, while the other end is combed with a single row of needles called the **top comb**.

Fig 3.27 shows the carding schematic diagram of cylinder and top comb in combed frame.

4. After both ends of the beard have been combed separately, the separated beards are then **reunited** by a **piecing** system.

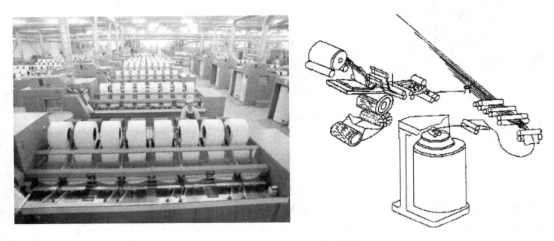

Fig 3.26　Combed frame(精梳机)

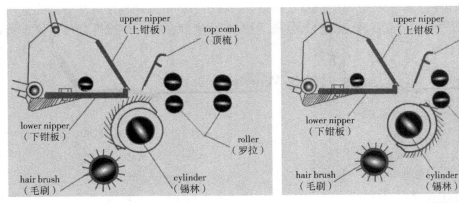

(a) Cylinder combing(锡林梳理)　　　　(b) Top combing(顶梳梳理)

Fig 3.27　Carding schematic diagram of cylinder and top combing in combed frame
(精梳机锡林与顶梳的梳理原理图)

Part 2 *Trying*

1. Purchase the same fineness of carded yarn and combed yarn from the market and compare them in appearance and performance.

	Carded Yarn	Combed Yarn
Appearance		
Performance		

2. Compare the surface and internal structure of carded yarn and combed yarn by drawing method (绘图法).

Carded Yarn	Combed Yarn

3. Under the guidance of the training teacher, learn the operation of carding machine in the training factory.

Part 3 *Thinking*

1. Why does the carding machine have a mixing effect on cotton fiber?

2. When you purchased the textiles in everyday life, have you encountered textile trademarks marked with combed yarn? What are the main textile products? What are the superior performance features of these products?

Task Four　Drawing and Roving Technology
（并条与粗纱技术）

【Text】

　　按照纱线的最终要求，棉条经梳棉、精梳工序，到并条工序[1]。并条工序提高了纤维的平行度，将几根生条并合成一根熟条[2]。这是个混合工序，有助于纱线均匀度的提高。并条工序是在并条机上完成的。其主要作用是去除生条缺陷，制成熟条[3]，这个工序是非常重要的，因为并条能大大提高最后成纱的强度和质量[4]。通常，6根或8根生条喂入并条机。这些生条经四对牵伸罗拉的作用，集束并牵伸至一根熟条[5]。对比熟条和最终的成纱，可见，前者比后者大约粗200倍。因为没有哪个传统的细纱机能通过一个工序完成这个任务，所以在并条和细纱工序之间，引入了粗纱工序。

【Key Words】

ultimate [ˈʌltimət] 最终的　　　　　　　　drawing frame　并条机
drawing [ˈdrɔːiŋ] 并条　　　　　　　　　　eliminate [iˈlimineit] 除去
parallelism [ˈpærəlelizəm] 平行性，平行度　attainment [əˈteinmənt] 达到
combine　联合，组合　　　　　　　　　　drawing roller　牵伸罗拉
contribute to　有助于　　　　　　　　　　dimension [daiˈmenʃn] 尺寸

【Key Sentences】

1. Depending on the **ultimate** yarn desired, slivers are processed through the carding unit or the combing unit to the **drawing**.

Fig 3.28 shows the drawing frame and its schematic.

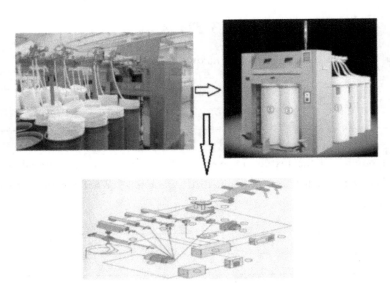

（a）Drawing frame（并条机）

Project Three Yarns and Spinning Technology(纱线与纺纱技术)

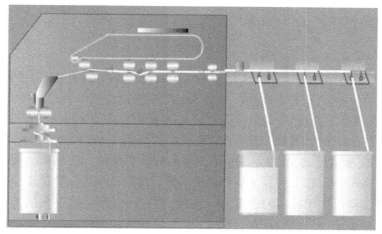

(b) Structure diagram of drawin frame(并条机结构简图)

Fig 3.28 Drawing frame and its schematic(并条机及其结构简图)

2. Drawing increases the **parallelism** of the fibers and combines several carded slivers into one **drawn sliver**.

Fig 3.29 shows the drawing and condensing part.

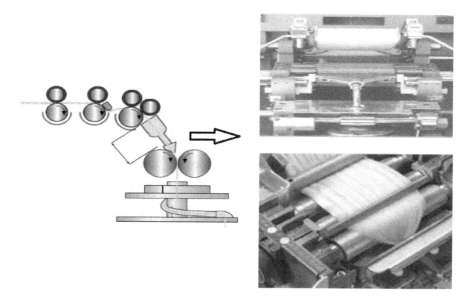

Fig 3.29 Drawing and condensing part(牵伸凝聚部分)

3. The functions of **drawing frame** are primarily to **eliminate** the defects from carded sliver and to produce a draw sliver.

4. The **attainment** of these objects is very important because it greatly improves the strength and general quality of the finally spun yarn.

5. They are condensed by the action of four pairs of **drawing roller**s, and drawn out to the **dimension** of one draw sliver.

【Text】

粗纱工序是一个从熟条到粗纱的工艺过程,简单地说,粗纱工序的作用是牵伸、加捻和卷绕[1]。粗纱牵伸是由3~4对罗拉实现的。粗纱机条筒中的生条被喂入三对牵伸罗拉之间,每对紧接着的罗拉要比前一对罗拉快,这对生条牵伸并将生条变薄,使纤维近乎伸直平行[2]。

粗纱通过锭子上的锭翼(转动)获得了少量的捻度[3]。锭子使得锭翼旋转并被驱动在一个恒定的速度。前罗拉(最接近锭翼)设定在一个速度从而赋予从前罗拉出来的须条预先确定的每英寸捻回数,当须条沿着罗拉和锭翼之间移动时[4],铜管的驱动来源与驱动锭子及锭翼的齿轮相互独立[5]。铜管被控制以快于锭翼的速度自动旋转,这可以引起粗纱以相同的速度卷绕在筒管上[6]。粗纱卷绕在筒管上,粗纱卷装两端呈锥台形[7]。这些工作的完成均依靠差微装置和铁炮来完成。

【Key Words】

roving ['rəuviŋ] 粗纱
winding ['waindiŋ] 卷绕
can 条筒
thin down 把…变薄
rubber roller 胶辊
rubber ring 胶圈
insert into 把…插入(某处)
flyer ['flaiə(r)] 锭翼
spindle ['spindl] 锭子
predetermined [ˌpriːdiˈtəːmind] 预先确定的,预先决定的
bobbin ['bɔbin] 铜管,粗纱管
gear [giə(r)] 齿轮,传统装置
regulate ['regjuleit] 调节,控制
automatically [ˌɔːtəˈmætikli] 自动地
sufficiently [səˈfiʃ(ə)ntli] 足以,十分,充分地
tapered ends 锥形端部
differential motion 差微装置
cone drum 铁炮

【Key Sentences】

1. **Roving** is an operation from drawing sliver to roving. Briefly, the functions of the roving process are drafting, twisting and **winding.**

Fig 3.30 shows the schematic of roving and roving frame.

2. The **can** of sliver from drawing frames is fed between three sets of drafting rollers. Each following set rollers runs faster than preceding set. This draws sliver and **thin**s it **down**, making fibers nearly parallel.

Fig 3.31 shows several different forms of drawing, Fig 3.32 shows the **rubber roller** and **rubber ring** of drawing frame.

3. A slight amount of twist is **inserted** into the strand by a **flyer** carried on the **spindle**.

4. The front rollers (nearest flyer) are set a speed that gives strand coming out of the rollers a **predetermined** number of turns of twist per inch as it moves along between rollers and flyer.

Project Three　Yarns and Spinning Technology(纱线与纺纱技术)

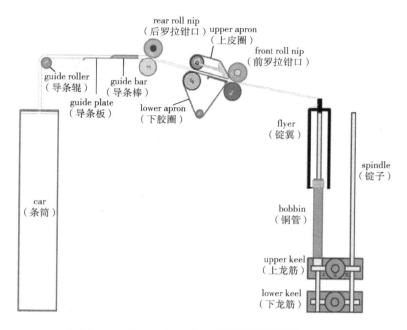

(a) Structure diagram of rovin frame (粗纱机结构简图)

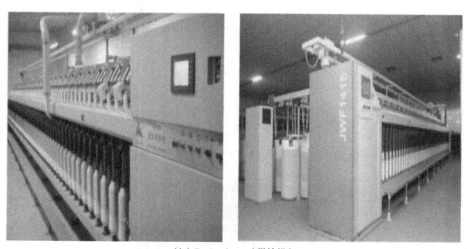

(b) Roving frame (粗纱机)

Fig 3.30　Schematic of roving and roving frame(粗纱机结构简图与粗纱机)

5. The **bobbin** is driven by a source separating from **gear** that drives spindle and flyer.

6. The bobbin is **regulated** to turn **automatically** at a speed **sufficiently** faster than flyer, which causes roving to wind on bobbin at the same speed.

7. Roving is laid onto a bobbin, and the package of roving has **tapered ends**.

Fig 3.33 shows the roving and Fig 3.34 shows the flyer.

99

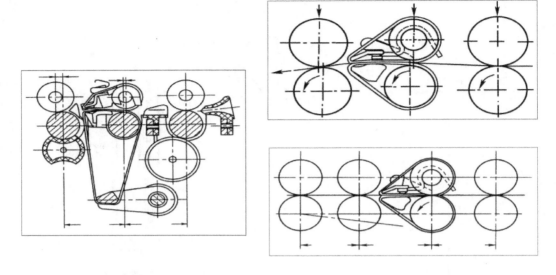

Fig 3.31　Several different forms of drawing(几种不同的牵伸形式)

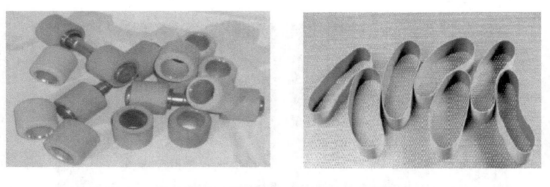

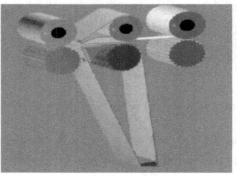

Fig 3.32　Rubber roller and rubber ring(胶辊与胶圈)

Project Three Yarns and Spinning Technology(纱线与纺纱技术)

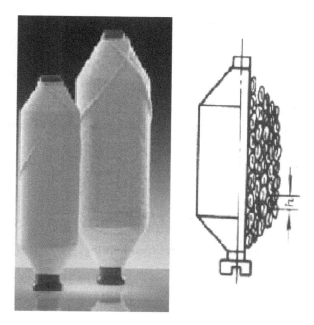

Fig 3.33 Roving(粗纱)

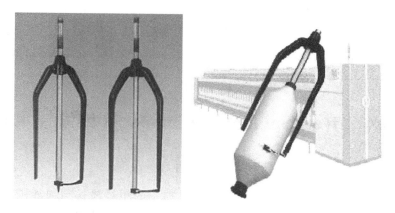

Fig 3.34 Flyer(锭翼)

Part 2 *Trying*

1. Please indicate that the change of the structure about carded sliver processed by drawing and roving using pictures.

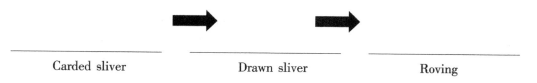

Carded sliver　　　　　　　Drawn sliver　　　　　　　Roving

2. Please represent why the drawing process can contribute to a greater yarn uniform by pictures.

3. Under the guidance of the training teacher, learn the operation of drawing frame and roving frame in the training factory.

Part 3 *Thinking*

1. What are the similarities and differences between drawing and roving process?

2. Why drawing process can achieve the mixing effect?

Task Five Spinning Technology(细纱技术)

【Text】

细纱工序是单纱生产的最后一道工序,粗纱机加工的粗纱输送到细纱机上纺制成纱线[1]。细纱工序的主要任务和粗纱一样,即牵伸、加捻和卷绕。牵伸就是将喂入的粗纱抽长拉细到所纺细纱规定的线密度[2]。这是纺纱准备的最后一道牵伸步骤。牵伸是由牵伸机构完成的,牵伸机构主要包括牵伸罗拉、罗拉轴承、胶辊、胶圈等[3]。

粗纱管粗纱被喂入几对牵伸罗拉对须条进行牵伸至最后需要的尺寸[4]。锭子使铜管以恒定的速度旋转。当须条经过时,前罗拉被调整以充足的速度传递纱线以施加所需的捻度[5]。

【Key Words】

single yarn 单纱
spinning frame 细纱机
attenuation [ə,tenjuˈeiʃn] 变薄;弄细
drafting mechanism 牵伸机构
drafting roller 牵伸罗拉
roller bearing 罗拉轴承
top roller 胶辊
apron [ˈeiprən] 胶圈
double apron 双皮圈
refined [riˈfaind] 精细的,精致的

Project Three Yarns and Spinning Technology(纱线与纺纱技术)

draw…down 把…牵伸至
glide [glaid] 滑动
drag [dræg] 拖,拉

【Key Sentences】

1. The spinning operation is the final process in the spinning of **single yarn**, and roving produced by the flyer frame is delivered to the spinning frame for spinning into yarn.

Fig 3.35 shows the spinning frame and its schematic. Fig 3.36 shows the main components of the spinning frame.

2. Drafting is the process of **attenuation** of the fed material to the required yarn with desired linear density.

3. Drafting is carried out by drafting mechanism which mainly includes drafting rollers, **roller bearings**, **top rollers**, **aprons** and so on.

Fig 3.37 shows the drawing system and its parts. Fig 3.38 shows the **double apron** on the drafting system.

4. Bobbin roving is fed between sets of drafting rolls to **draw** strand **down** to its final desired size.

5. The front set of rolls is adjusted to deliver yarns at a speed sufficient to insert desired amount of twist as strand moves along.

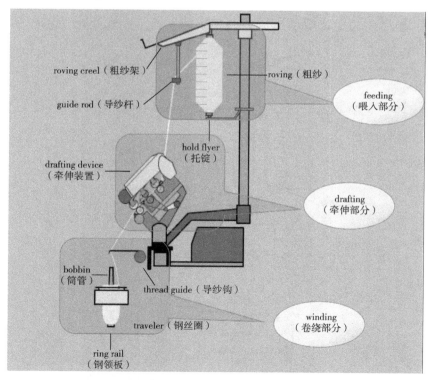

(a) Structure diagram of spinning frame (细纱结构简图)

Fig 3.35

103

(b) Spinning frame（细纱机）

Fig 3.35　Spinning frame and its chematic（细纱机及其结构简图）

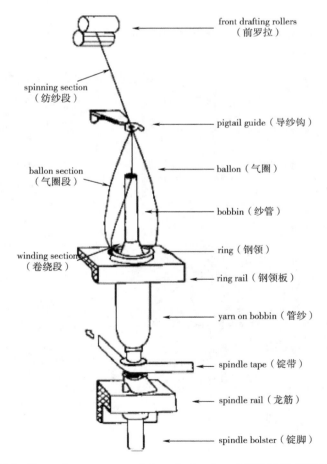

Fig 3.36　The main components of the spinning frame（细纱机的主要组成部分）

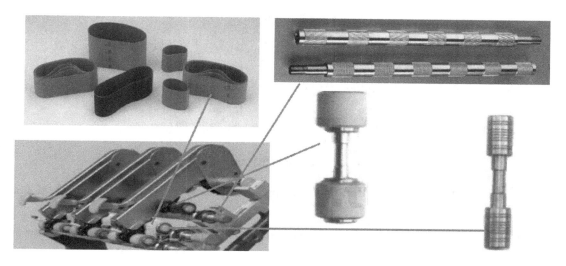

Fig 3.37　Drawing system and its parts(牵伸系统及部件)

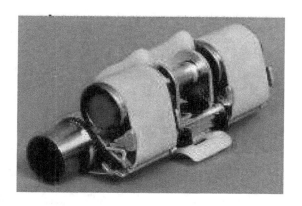

Fig 3.38　Double apron drafting system(双胶圈牵伸系统)

【Text】

　　加捻就是在抽长拉细的纤维束中加上捻度,使纤维增加抱合力[1]。加捻是为了使纱线具有足够的强度以便于制成织物[2]。加捻数由纱的线密度、纤维性能及纱线用途来决定。所谓卷绕,就是纱线卷绕到筒管上,以形成合适的卷装[3]。实际上,卷绕不是和加捻分开的,它们由同台设备同时完成。

　　细纱机加捻卷绕的主要元件有锭子、筒管、钢领、钢丝圈等。钢领和钢丝圈是细纱机上最重要的部分[4]。须条由前罗拉输出,通过导纱钩、钢丝圈,卷绕到安装在锭子上的筒管上,并随锭子一起旋转[5]。钢丝圈在钢领上转一圈便给细纱加上一个捻回。钢丝圈围绕钢领自由滑动,当纱线被罗拉传递时,由钢丝圈牵拉引起的张力使其以相同的速率卷绕在细纱管上[6]。所以,除了加工更加精细,且用钢领与钢丝圈来替代锭翼外,细纱原理与粗纱原理相同[7]。

【Key Words】

attenuate [əˈtenjueit] 变细,变薄
insertion [inˈsə:ʃn] 插入
sufficient [səˈfiʃnt] 足够的,充足的,充分的
be fabricated [ˈfæbrikeitid] into 被制成
bobbin [ˈbɔbin] 细纱筒管
package [ˈpækidʒ] ring 钢领
traveler 钢丝圈
spinning frame 细纱机
thread guide 导纱钩
spindle 锭子

【Key Sentences】

1. Twisting is the process of **insertion** of twist into the **attenuated** fiber bundle to bind the fibers together.

2. Twisting is done in order to give **sufficient** strength to enable yarn to **be fabricated into** fabric.

3. Winding is to wind the spun yarn onto a **bobbin** to produce a suitable **package**.

4. On the spinning frame, **ring** and **traveler** are the most important parts of **spinning frame**. Fig 3.39 shows ring, traveler and how they are arranged.

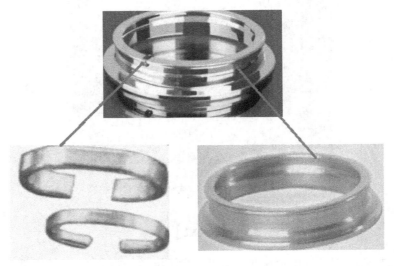

Fig 3.39 Ring and traveller(钢领与钢丝圈)

5. When the strand is delivered by the front rollers, it passes through the **thread guide**, traveler and wind onto the bobbin which is placed on the spindle and rotats together with the spindle.

Fig 3.40 shows the spindle. Fig 3.41 shows the thread guide.

6. The traveler **glides** freely around the ring. The tension caused by **drag** of traveler causes yarn to wind on bobbin at same rate of speed as it is delivered by rolls.

Fig 3.42 shows the bobbin on the spinning frame.

7. The principle of spinning is the same as that of roving except that the operation is more **refined** and a ring and traveler are used instead of the flyer.

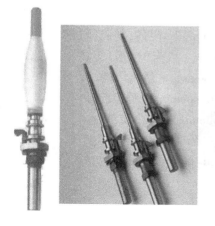

Fig 3.40　Spindle(锭子)

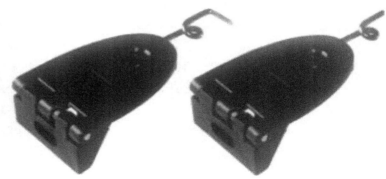

Fig 3.41　Thread guide(导纱钩)

Fig 3.42　Bobbin(筒管)

【Text】

由于环锭纺纱机具有加捻和卷绕是有机整体的内在属性[1]。过小卷装和低纺纱速度是解决卷装和卷绕矛盾的瓶颈,也阻碍了高纺纱速度、大卷装以及自动化工序的实现[2]。

现在,市场上出现了许多新型纺纱技术,如气流(转杯)纺纱、涡流纺纱、静电纺纱,以及摩擦纺纱等[3]。新型纺纱与环锭纺纱最大的区别在于将加捻与卷绕分开进行,使加捻与卷绕互不牵制[4]。因此,新型纺纱具有产量高、卷装大(筒纱)、流程短(条子直接喂入,筒纱输出)等优点[5]。

【Key Words】

intrinsic [inˈtrinsik] 固有的,内在的,本质的
attribute [əˈtribjuːt] 属性,特征
organic [ɔːˈɡænik] whole 有机整体
bottleneck [ˈbɔtlnek] （工商业发展）瓶颈,阻碍
contradictory [ˌkɔntrəˈdiktəri] 对立物,矛盾因素
impede [imˈpiːd] 阻碍,妨碍,阻止
rotor spinning 气流(转杯)纺纱
air votex spinning 涡流纺纱
electrostatic spinning 静电纺纱
friction spinning 摩擦纺纱
hold back 阻碍
yield [jiːld] 产量,产额

【Key Sentences】

1. Because ring spinning machine possesses **intrinsic attribute** of an **organic whole** for twisting and winding.

2. Very small package and low spinning speed are its **bottleneck** to solve the **contradictory** of package and winding. It also **impedes** the realization of high spinning speed and big package and automation as well.

3. Now there are many new spinning technology on the market, such as **rotor spinning**, **air votex**, **electrostatic spinning**, as well as **friction spinning** and so on.

Fig 3.43 shows the rotor spinning frame, Fig 3.44 shows the air votex spinning frame. Fig 3.45 shows the schematic of electrostatic spinning. Fig 3.46 shows the friction spinning and its schematic.

Fig 3.43　Rotor spinning frame(转杯纺纱机)

Project Three Yarns and Spinning Technology(纱线与纺纱技术)

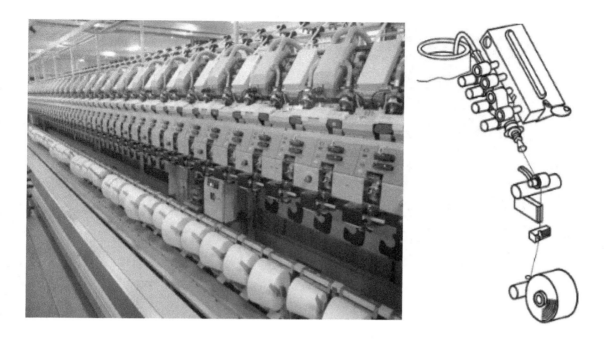

Fig 3.44 Air votex spinning frame(涡流纺纱机)

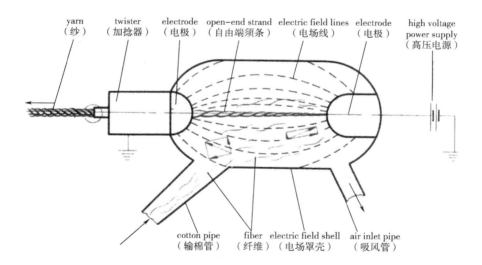

Fig 3.45 Schematic of electrostatic spinning(静电纺纱)

4. The biggest difference between the new spinning and the ring spinning is that the twisting and winding are separated, so that the twisting and winding are not **held back** each other.

5. The new spinning has the advantages of high **yield**, large package (cone), short process (direct feeding of sliver, output of cone) and so on.

109

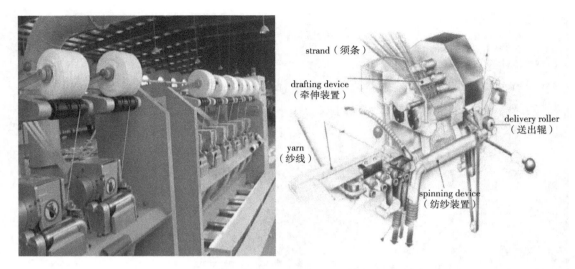

Fig 3.46 Friction spinning and its schematic(摩擦纺纱及其结构简图)

Part 2 *Trying*

1. Buy the same raw materials, the same line density of the ring spinning yarn and rotor spinning yarn from the market as much as possible, describe their differences on structure and performance.

	Ring spinning yarn	Rotor spinning yarn
Structure		
Performance		

2. Combined with the previous knowledge you learned, please briefly write the spinning process of the traditional carded ring spinning yarn and air votex spinning yarn.

a. Traditional carded ring spinning yarn

_____
_____
_____
_____

b. Air votex spinning yarn

_____
_____
_____

3. Under the guidance of the training teacher, learn the operation of spinning frame in the training factory.

Part 3 *Thinking*

1. Combined with daily life experience, which clothing fabrics do you think use ring spinning yarn? Which clothing fabrics use the new spinning yarn?

_____
_____
_____
_____

2. What opportunities and challenges will the new spinning technology bring to spinning enterprises?

_____
_____
_____
_____

Task Six　Winding Technology(络筒技术)

【Text】

（纺纱）加工后卷装而成的纱线不在最佳条件下不能用于织制织物。成纱后,短纤纱和连续长丝纱都不能立即用于织物成型系统[1]。卷装尺寸、结构和其他因素使纱线有必要进行进一步加工,在织物形成过程中以便纱线被有效地处理[2]。

对机织和经编来说,多根纱线同时以经纱片的形式呈现[3]。这些纱线取自于卷装的织轴上。有梭织机需要特殊的纬纱卷装或纡管,纡管适用于梭子,而无梭织机或纬编机使用扁圆形筒子或锥形筒子大卷装上的纱线[4]。由此可知,细纱卷装的纱线实际上是无用的。纱线需要重新卷装达到织物成型系统的特殊需求[5]。事实上,这是纱线准备的其中一个功能,就是为了特殊的织物成型系统将纱线置于合适的卷装上。

因此,对于机织和针织系统来说,纱线准备的第一步都是络筒[6]。络筒的目的主要有两点:形成能够满足后序加工的卷装;检验和清洁纱线(比如除去粗节和细节)[7]。为完成上述两个任务,络筒机分为三个区域:退绕区;张力和清洁区;卷绕区[8]。

【Key Words】

fabric forming systems　织物成形系统
package size　卷绕尺寸
warp knitting　经编
simultaneously [ˌsɪməlˈteɪnɪəsli]　同时
warp sheet　经纱片
beam　织轴
shuttle loom　有梭织机
quill [kwɪl]　纡管
shuttleless loom　无梭织机
cheese　扁圆柱形筒子
cone　圆锥形筒子
repackage [ˌriːˈpækɪdʒ]　重新包装
weaving process　机织工艺
knitting process　针织工艺
winding [ˈwaɪndɪŋ]　络筒
suitable [ˈsuːtəbl]　合适的,适当的
thick and thin spots　粗细节
winder [ˈwaɪndə(r)]　络筒机
unwinding zone　退绕区
automatic winder　自动络筒机

【Key Sentences】

1. After yarn formation, both spun and continuous filament yarns are not immediately usable in **fabric forming systems**.

2. **Package size**, structure and other factors make it necessary for the yarns to be further processed to prepare them to be handled efficiently during fabric formation.

3. For weaving and **warp knitting**, many yarns are presented **simultaneously** in the form of a **warp sheet**.

4. **Shuttle looms** need a special filling yarn package, or **quill**, which fits in the shuttle; while **shuttleless looms** and weft knitting machines use yarn from large packages called **cheeses** or **cones**.

5. The yarn must be **repackaged** to meet the particular needs and demands of the fabric forming system in which it is to be used.

6. Therefore, for both **weaving and knitting processes**, the first step in yarn preparation is **winding**.

Fig 3.47 shows the winding (changing packages from bobbin to cone).

7. The reasons for winding yarns are as follows: winding can produce a package which is **suitable** for further processing, winding can inspect and clean the yarns.

Fig 3.48 shows thick spot, thin spot, slub and nep.

8. In order to carry out the above two tasks, a **winder** is divided into three principal zones: the **unwinding zone**, the tension and clearing zone, the winding zone.

Fig 3.49 shows different kinds of **automatic winders** and Fig 3.50 shows three principal zones of the automatic winder.

【Text】

要将纱线重新卷绕在新的卷装上,必须首先从旧卷装上移除[1]。这个任务是在退绕区完成的。这个区域仅包含一个纱架,该纱架将原始的卷装(细纱管纱)放置在合适的位置方便退绕[2]。

Project Three　Yarns and Spinning Technology(纱线与纺纱技术)

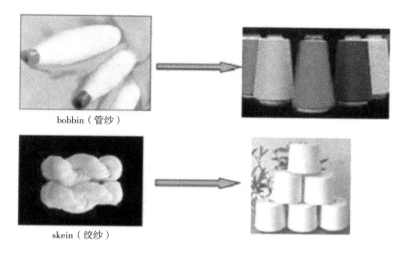

Fig 3.47　Winding(from bobbin to cone)(络筒)

Fig 3.48　Thick spot, thin spot, slub and nep(粗节、细节、竹节与棉节)

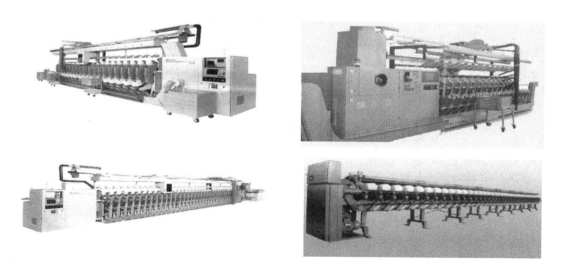

Fig 3.49　Automatic winder(自动络筒机)

113

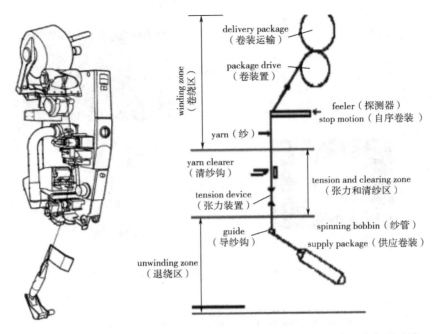

Fig 3.50　Three principal zones of the automatic winder(自动络筒机的三个主要区域)

常用的退绕方法有侧向退绕和轴向退绕。侧向退绕时,两端带边盘的纱管必须经旋转才能将纱线退下来[3]。该系统的优点是纱线在退绕时不会旋转,因此,纱线捻度保持不变[4]。它的缺点是,筒管必须旋转,在高卷绕速度下,由于惯性,筒管的旋转可能会导致纱线的张力变化[5]。使用轴向退绕时,纱线是从卷装的一端抽拉出来的,因此,卷装本身并不需要旋转[6]。这是最简单也是最普遍的纱线退绕方式。

【Key Words】

rewind ［ˌriːˈwaind］ 重绕,倒回　　　　over-end withdrawal　轴向退绕
creel ［kriːl］ 纱架,筒子架　　　　　　flanged ［flændʒd］ 带边缘的
side withdrawal　侧向退绕　　　　　　 inertia ［iˈnəːʃə］ 惯性

【Key Sentences】

1. To **rewind** the yarn on a new package, it must first be removed from the old package.

2. The winding zone merely consists of a **creel**, which holds the old package in an optimum position for unwinding.

3. In the **side withdrawal**, the **flanged** bobbin must rotate in order for the yarn to be removed. Fig 3.51(a) shows the side withdrawal.

4. The advantage of the side withdrawal is that the yarn does not rotate upon withdrawal and therefore the yarn twist remains constant.

5. At high winding speeds, due to **inertia**, the rotation of the bobbin may lead to tension varia-

tions in the yarn.

6. In the **over-end withdrawal** method, the packages don't need to rotate as the yarn is pulled over the end of the package.

Fig 3.51(b) shows the over-end withdrawal.

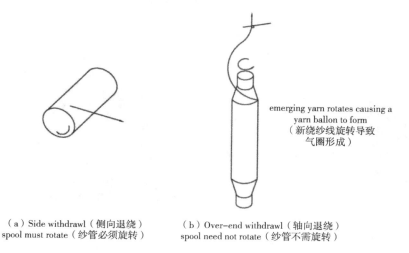

(a) Side withdrawl（侧向退绕） spool must rotate（纱管必须旋转）　　(b) Over-end withdrawl（轴向退绕） spool need not rotate（纱管不需旋转）

Fig 3.51　Side withdrawal and over-end withdrawal(侧向退绕与轴向退绕)

【Text】

下一个区域是张力和清洁区。在这个区域，纱线会接受适当的张力，既是为了提供适当的卷装密度，也是为了方便进一步的加工¹。这个区域包括张力装置，它可保持纱线的适当张力，以实现均匀的卷装密度；包括清纱器，它是用于检测纱线上的粗节和细节，并清除掉它们；还包括自停装置，当纱线断头或者卷装用尽的时候，它可以使卷绕停止²。

最后，纱线在卷绕区卷绕在合适的卷装上。卷装有各种形式，可以是圆锥形筒子、扁圆形筒子、有边筒子，这取决于下道工序的要求³。重要的是，在卷绕的时候，不会发生捻度变化。因此，在卷绕过程中，应避免筒子不动将纱线进行缠绕，纱线仅仅是靠筒子的转动而发生缠绕⁴。

【Key Words】

tension and clearing zone　张力和清洁区
tension device　张力装置
maintenance[ˈmeɪntənəns]　维持，保持
clearer　清纱器，清除器
automatic stop motion　自停装置
in the case of　至于…，就…来说

depletion [dɪˈpliːʃən] 用尽，耗尽
cone [kəʊn] 圆锥形
cheese [tʃiːz] 扁圆形筒子
spool [spuːl]（空心而两端有突缘的）有边筒子
physically [ˈfɪzɪkli] 自身地

【Key Sentences】

1. It is in the tension and clearing zone that the yarn receives the proper tension to provide an ac-

ceptable package density and build for further processing.

2. The tension and clearing zone consists ofa **tension device** which allows the **maintenance** of proper tension in the yarn in order to achieve a uniform package density, a **clearer** which detects thick or thin spots on the yarn and clears them, and an **automatic stop motion** which causes the winding to stop **in the case of** a yarn break or the **depletion** of a supply package.

Fig 3.52 shows the electronic clearer.

Fig 3.52　Electronic clearer(电子清纱器)

3. The package may be one of many types, a **cone**, a **cheese**, a **spool**, depending on the next operation.

Fig 3.53 shows the cylindrical. Fig 3.54 shows the cone. Fig 3.55 shows bobbins of other shapes.

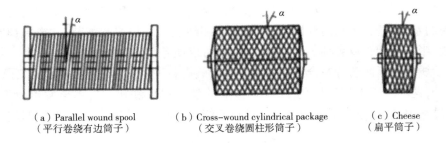

(a) Parallel wound spool　　　(b) Cross-wound cylindrical package　　　(c) Cheese
（平行卷绕有边筒子）　　　　（交叉卷绕圆柱形筒子）　　　　　　（扁平筒子）

Fig 3.53　Cylindrical(圆柱形筒子)

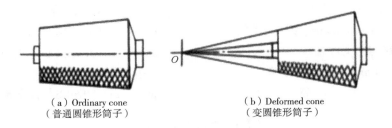

(a) Ordinary cone　　　　　(b) Deformed cone
（普通圆锥形筒子）　　　　（变圆锥形筒子）

Fig 3.54　Cone(圆锥形筒子)

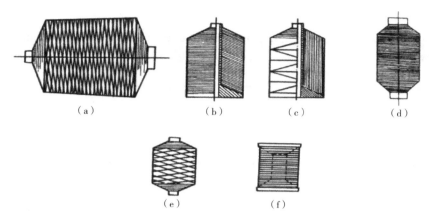

Fig 3.55 Bobbins of other shapes(其他形状的筒子)

4. Thus **physically** wrapping the yarn around the package during winding should be avoided and the yarn is wound on the package by only rotating the package.

Part 2 *Trying*

1. Seek a kind of ring spinning yarn after the winding process from the local spinning enterprises, to analyze and compare with their structure and performance.

	Ring spinning yarn	Ring spinning yarn after winding
Structure		
Performance		

2. Combined with the previous knowledge you learned in this course, write the process from cotton fiber to cone yarn (cotton fiber need to be combed).

_____
_____
_____
_____

3. Under the guidance of the training teacher, learn the operation of automatic winder in the training factory.

Part 3 *Thinking*

1. New spinning such as rotor spinning and air votex spinning, do they need carry out the winding process? Why?

2. From the economic point of view, what do you think are the advantages and disadvantages of the winding process?

Project Four Woven Fabrics and Woven Technology(机织物与机织技术)

Task One Warping Technology(整经技术)

【Text】

如果织物成型系统是机织或针织,则部分或全部纱线以片纱形式呈现后形成织物[1]。因此,有必要将络筒纱从络筒卷装上退绕下来,并且按一定的数目要求卷绕在经轴上[2]。纱线必须平行排列,并且在张力均匀的状态下,以及上述提到的纱线按所设计的排序来整理[3]。这些工作就是整经,整经是将纱线卷绕到经轴上的工序,为浆纱做准备[4]。

【Key Words】

warp knitting 经编
sheet form 片纱形式
uniform tension 均匀的张力
beam 经轴,织轴,卷轴

uniform tension 均匀的张力
sequence 顺序,序列
warping [ˈwɔːpiŋ] 整经
slashing [ˈslæʃiŋ] 浆纱,经纱上浆

【Key Sentences】

1. If the fabric forming system is weaving or **warp knitting**, some or all the yarns forming the fabric are presented in **sheet form**.

2. It is necessary to remove the yarns from the winding package and arrange the desired number on a package called **beam**.

3. The yarns must be parallel and under **uniform tension** and arrange the above mentioned threads according to the desired **sequence**.

4. **Warping** is the operation of winding warp yarns onto a beam usually preparation for slashing. Fig 4.1 shows the schematic of warping.

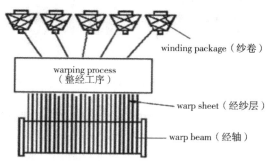

Fig 4.1 Schematic of Warping(整经)

【Text】

　　设立一个架子来存放络筒卷装是合理的,这就是所谓的筒子架,它的主要功能就是以一定的方式放置供应的卷装以便整经[1]。为了顺利地完成整经,筒子架上安装了锭座,用来放置供纱卷装,便于纱线退绕[2]。筒子架上还装备如下部件:张力装置用来帮助维持整个筒子架上所有的纱线张力均匀;引导器用来引导纱线并帮助保持经纱分隔开[3];防静电装置用来消除纱线摩擦各种表面所产生的静电[4];自停装置,用来检测断经或卷装用尽[5]。

　　筒子架容量,即筒子架上所能容纳的筒子的数量,是整经工艺一个非常重要的因素,因为它是计算整经条数和经轴数时需要依据的参数[6]。总的来讲,最大的筒子架容量范围为粗支纱的300个筒子到细支纱的1400个筒子[7]。

【Key Words】

logical ['lɔdʒikl] 合理的,复合逻辑的
frame 框,架
creel [kri:l] 筒子架
facilitate [fə'siliteit] 使容易
package holders 筒子锭座
tension device 张力装置
maintain [mein'tein] 保持,维持
guide 名词,可译为引导器

direct 动词用,引导
keep apart (使)分开
ends 可译为经纱
antistatic [ˌænti'stætik] 抗静电的
static charges ['stætik tʃɑːdʒ] 静电荷
stop motion 自停装置
parameter [pə'ræmitə(r)] 参数

【Key Sentences】

1.It is **logical** to build a **frame** for holding the winding packages, and this frame is known as a **creel** and its function is to hold the supply packages in a manner so as to **facilitate** warping.

Fig 4.2 shows the creel.

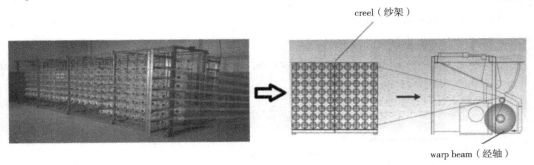

Fig 4.2　Creel（筒子架）

2. To accomplish the purpose, creels are equipped with **package holders** on which the supply packages are placed, easy for unwinding.

3. The **tension devices** are used to help **maintain** uniform tension throughout the creel, **guides** to **direct** the yarn and to help **keep** the **ends apart**.

Project Four Woven Fabrics and Woven Technology(机织物与机织技术)

4. The **antistatic** devices are used to eliminate **static charges** created by the rubbing of the yarn against the various surfaces.

5. The stop motions are used to detect broken ends or empty packages.

6. Creel capacity, which is the number it may hold, is an important factor in warping, because it is the **parameter** on which the number of warping sections or beams depends.

7. In general, maximum creel capacity ranges from about 300 packages for very high yarns to 1400 packages for thin yarns.

【Text】

整经取决于筒子架的容量、经轴上最终需要的经纱根数以及是否需要在经纱上排列花型等多种因素,例如,织物上的经向条纹[1]。在纺织工业上,整经工序可按两种不同的工艺来进行:分条整经和分批整经[2]。如果筒子架容量足够大,所需的经纱总数足够低,或者,如果筒子架容量不足以提供所需的经纱数,并且不需要明显的纱线花纹,则通常使用分批整经[3]。分批整经是简单而直接地将供应卷装上的纱线卷绕到经轴上。

【Key Words】

pattern [ˈpætn] 花样,图案
sectional [ˈsekʃənl] warping = drum warping 分条整经
beam warping 分批整经
sufficiently [səˈfiʃntli] 足够地,充分地
distinct [diˈstiŋkt] 明显的,清楚的

【Key Sentences】

1. The warping is done depends on the capacity of the creel, the number of ends required in the final beam and the necessity of maintaining a **pattern** in the warp, for example, the warp strips in the fabric.

2. The industrial warping process can be carried out according to two different methods: **sectional warping** and **beam warping**.

3. If the creel capacity is **sufficiently** high and the total number of ends required is sufficiently low or, if creel capacity is not sufficiently to supply all the required ends and no **distinct** yarn pattern is required, then **beam warping** is generally used.

Fig 4.3 shows the beam warping machine. Fig 4.4 shows the warp dressing in beam warping.

Fig 4.3 Beam warping machine(分批整经机)

121

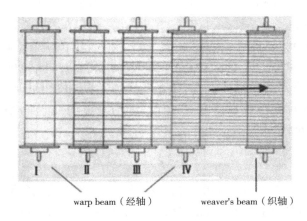

warp beam（经轴）　　weaver's beam（织轴）

Fig 4.4　Warp dressing in beam warping(经纱并轴)

【Text】

如果没有足够的筒子架容量，有必要构建一个含有所需全部经纱的整经轴，或者如果经纱必须按一定的顺序排列，则使用分条整经[1]。在分条整经中，经纱不是直接从筒子架到经轴上，而是经纱部分地被卷绕到大圆框上[2]。在这种方式中，所有的经纱以系列条带建立在大圆框上[3]。当织物所需的经纱总数被缠绕在大滚筒上时，它们都会同时被移除并卷绕在织轴上[4]。这个织轴包含经纱所需的确切经纱数。当经纱从筒子架上取出，并卷绕在大滚筒上时，经纱相互之间精确的位置已经确定[5]。

【Key Words】

insufficient [ˌinsə'fiʃnt] 不足的，不够的
totality [təu'tæləti] 整体，全部，总数，总额
definite ['definət] 明确的，一定的
not…but rather 不是…而是
build up 建造，建立

section 条带
pattern drum 大圆框，此处可称为大滚筒
simultaneously [ˌsiməl'teiniəsli] 同时地
take from 取自…
placement 定位，位置

【Key Sentences】

1. If, however, with **insufficient** creel capacity, it is necessary to build a warp beam containing the **totality** of ends required, or if the warp yarns have to be arranged in a **definite** order, then sectional warping is used.

Fig 4.5 shows the sectional warping machine. Fig 4.6 shows the inverted shaft mechanism of sectional warping machine.

2. In sectional warping, the warp is **not** wound directly from the creel onto the beam **but rather** sections of the warp are wound onto a pattern drum.

3. In this manner, the entire warp is **built up** in a series of sections on the pattern drum.

Project Four Woven Fabrics and Woven Technology(机织物与机织技术)

Fig4.5 Sectional warping machine(分条整经机)

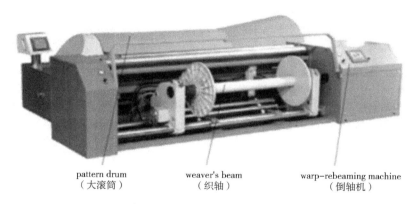

pattern drum　　　weaver's beam　　　warp-rebeaming machine
（大滚筒）　　　　（织轴）　　　　　（倒轴机）

Fig 4.6 Inverted shaft mechanism of sectional warping machine(分条整经的倒轴机构)

Fig 4.7 shows the fabrics suitable for processing are suitable for sectional warping and beam warping.

（a）Sectional warping（分批整经）　　　（b）Beam warping（分条整经）

Fig 4.7 Fabrics suitable for processing are suitable for sectional warping and beam warping
（分批整经与分条整经分别适合加工的织物）

4. When the total number of warp ends required in the fabric have been wound on the pattern drum, they are all removed **simultaneously** and wound upon a beam.

5. When the ends are **taken from** the creel and wound on the pattern drum, exact **placement** in relation to each other may be made.

Part 2 *Trying*

1. Please demonstrate the difference between sectional warping and beam warping by drawing.

Sectional warping
Beam warping

2. Find out the principle of sectional warping and beam warping, what kind of warping method should be used for pure cotton gray fabric and red and blue stripe fabric? Why?

_____
_____
_____
_____

Part 3 *Thinking*

1. Do you think what kind of textile fabrics seen in your daily life use sectional warping? Which kind of textile fabrics use beam warping? Give a few examples to illustrate.

_____
_____
_____
_____

2. Is there any difference of the yarn used in thesectional warping and the beam warping? What is the difference?

_____
_____
_____
_____

Task Two　Warp Sizing Technology(浆纱技术)

【Text】

经纱上浆总的目的就是生产能够承受机织过程中种种问题的经纱[1]。这些问题主要包括：与金属部件之间的摩擦，因为经纱需要穿过停经片、综丝、钢筘等金属部件；因为强力不够而产生的纱线断头；在开口过程中，相邻纱线相互移动所产生的毛羽缠结[2]。

经纱上浆的主要目的是加工在织造中遭受最小破坏的经纱。在某些情况下，浆纱也用于改变纱线的特性，以对织物的重量、刚度或手感产生影响，但如果第二个作用会阻碍第一个主要作用，那么浆纱工艺被误应用[3]。经纱上浆通过使纤维相互黏附的方式使纱线更坚固、光滑和润滑来达到主要目的[4]。尽管如此，浆料不应该阻碍织造的下道工序也是非常重要的。无论何时，浆纱应该有助于这些工序而不是阻碍它们[5]。因此，不仅要考虑浆料应用的方法以及对织造的影响，也要考虑对后续工艺(如染色)和织物带来的影响。

【Key Words】

slashing　浆纱
warp sizing　浆纱
withstand [wiðˈstænd]　经受，承受
rigor [ˈrigə]　艰苦，苛刻
drop wire　停经片
heddle [ˈhedl]　综丝
reed　钢筘
tangling　缠结，混乱

shedding [ˈʃediŋ]　开口
modify [ˈmɔdifai]　调整，改变
interfere with　阻碍
misapply [ˌmisəˈplai]　误用；滥用
mutually [ˈmjuːtʃuəli]　相互地，彼此
lubricated [ˈluːbrikeitid]　润滑的
wherever possible　尽可能
eliminate [iˈlimineit]　清楚，消除

【Key Sentences】

1. The total purpose of **slashing** or **warp sizing** is to produce a warp which will **withstand** the **rigors** of weaving.

Fig 4.8 shows the parts of yarn passing in weaving.

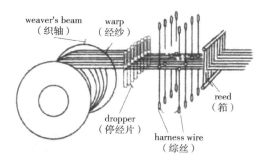

Fig 4.8　Parts of yarn passing in weaving(织造中纱线经过的部件)

2. The rigors mainly include: rubbing against metal parts by being threaded through **drop wires**, **heddles** and **reed**; yarn breakage because of strength is inadequate; **tangling** of hairiness on yarns which move back and forth with each other during **shedding**.

Fig 4.9 shows the hairiness of yarn.

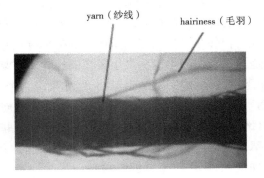

Fig 4.9　Hairiness of yarn(纱线毛羽)

3. In some cases, it is also used to **modify** the character of the yarn so as to have an effect on the fabric weight, stiffness or hand, but if this secondary use should **interfere with** the primary one, then the process has been **misapplied**.

4. Warp sizing achieves its primary purpose by causing fibers **mutually** to adhere in such a way as to make the warp yarns stronger, smoother and better **lubricated**.

Fig 4.10 shows the cross section of sizing yarn. Fig 4.11 shows the purpose of sizing.

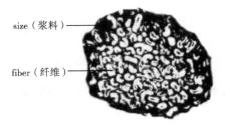

Fig 4.10　Cross section of sizing yarn(上浆后纱线的横截面)

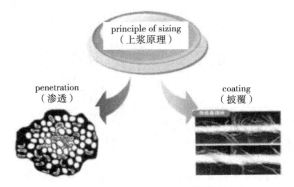

Fig 4.11　Purpose of sizning(浆纱目的)

5. **Wherever possible** the sizing should be helpful for these processes and not hinder them.

【Text】

虽然最终目标是要去除所有浆纱(以及其他准备工序)来达到最小成本,但在目前是不切实际的[1]。实际上,充分的准备不但可以使织造工序获得更高效率来降低总成本,而且大部分经纱在工业中不经浆纱进行织造是不可能的[2]。浆料通常应用于纱线以赋予它们额外的强度,但即使对于连续长丝,其强度是足够的,缺少浆料也许会允许松弛或断裂的长丝突出在纱体,特别是低捻纱和变形丝[3]。这些凸起的纱线能纠缠或起球引起纱线断头[4]。一般来说,捻度越高,需要的浆料就越少。如果捻度特别高,没有浆纱的经纱可进行织造,但由此形成的织物太粗糙且大多数目的用途不能适用[5]。使用未上浆的股线进行织造可以获得可被接受的织物,但合股的成本必须与浆纱的成本相互抵消[6]。

一般来讲,短纤纱必须上浆。连续长丝通常需要黏着剂来保护其避免断裂[7]。股线上浆主要是为了润滑或者使其表面光滑。浆纱的效果会受到很多因素的影响,主要包括浆料的配方和温度,设备操作状况,例如,经纱行进速度、烘燥温度和纱线张力等;纱线的吸浆量等。

【Key Words】

preparatory [pri'pærətri] 预备的;筹备的
minimum ['miniməm] 最低的,最小的
practicable ['præktikəbl] 可行的,行得通的
indeed [in'di:d] 实际上
overall cost 总成本
adequate ['ædikwət] 足够的,充分的
slack [slæk] 松弛的
protrude [prə'tru:d] 突出,伸出
textured yarn 变形丝,变形纱

fuzz balls 起球
exceptionally [ik'sepʃənəli] 特别,非常
harsh 粗糙的
set against 使…与…对立(矛盾),此处译为可抵消
protect……from 免于遭受……
lubrication [ˌlu:bri'keiʃn] 润滑
size recipe ['resəpi] 浆料配方
pick up 吸收

【Key Sentences】

1. Although the ultimate aim should be to **eliminate** all sizing (and other **preparatory** processes) to achieve **minimum** cost, this is not **practicable** at present.

2. **Indeed**, not only does adequate preparation reduce **overall costs** by making the weaving operation more efficient, but it is almost impossible to weave most warps found in industry without sizing.

3. Size is often applied to yarns to give them added strength, but even with continuous filaments where the strength is **adequate**, lack of size may allow **slack** or broken filaments to **protrude** from the body of the yarn, especially in the cases of low twist and **textured yarns**.

4. These protruding yarns can entangle or form **fuzz balls** and cause end breaks.

5. If the twist is **exceptionally** high, it is possible to weave a warp without size, but the resulting fabric is **harsh** and unacceptable for most purposes.

6. Acceptable fabrics may be obtained by weaving unsized ply yarns but the cost of plying has to be **set against** the cost of slashing.

7. Continuous filaments usually need adhesive to **protect** the filaments **from** breaking.

【Text】

如果吸浆过多,纱线会因浆液的重量而发脆,因此,会出现过多的经纱断头[1]。如果吸浆过少或者没有吸浆,那么上浆的效果就不会实现,同样会出现过多的经纱断头[2]。考虑未来的发展,新型的纱线结构可能会改变上浆需求。紧密纺纱线(与普通环锭纱相比,其毛羽较少)需要较少的上浆率,无捻纱(靠黏合剂或者特定的长丝使纤维抱合在一起)很显然需要很小的上浆率甚至不需要上浆,任何浆纱尝试都很有可能会造成无捻纱线的损坏,特别是如果原来的黏合剂是水溶性的[3]。

而且,织机也在不断的改进。有梭织机正在让位给无梭织机,施加在经纱上的条件也在改变。例如,喷水织机,它的特殊之处是使用水流引纬,如果使用可溶于水的浆料来上浆的话就会出现问题[4]。

【Key Words】

brittle ['britl] 易碎的
excessive [ik'sesiv] 过多的,过分的,过度的
end break 经纱断头
breakages ['breikidʒiz] 毁坏,损坏
modify ['mɔdifai] 修改,改变
compact [kəm'pækt] yarn 紧密纺纱

hairiness ['heərinəs] 毛羽
twistless yarn 无捻纱
waters jet loom 喷水织机
filling ['filiŋ] 纬纱
water-soluable ['wɔːtərsɔlj'ubl] 可溶于水的
give way to 让路给

【Key Sentences】

1. If the yarn contains too much size by weight, it will tend to be **brittle** and, as a result, an **excessive** number of **end breaks** will occur.

2. If the yarn contains little or no size, then none of the benefits of sizing will be realized and again there will be excessive end **breakages**.

3. **Compact yarns** (which have less **hairiness** than common ring spinning yarns) may need little sizing and **twistless yarns** (in which fibers are held together by an adhesive) obviously need little or no further sizing and it is quite possible that any attempt to slash them would cause yarn damage especially when the original adhesive were water soluble.

4. The **water jet loom**, for example, is a special case in which the **filling** is inserted by a jet of water and this obviously introduces sizing problems when a **water-soluable** size is used.

【Text】

一般来说,浆料成分可分为四类:黏合剂、润滑剂、助剂和溶剂[1]。一般情况下,浆纱机可分

为五个不同的部分:经轴架、浆槽、烘房、纱线分绞区和机头。经轴架只是一种框架装置,在框架上,经轴处于一个方便退绕的状态[2]。浆槽中包含浆纱溶液,称为浆液。纱线通过引导罗拉进入浆槽,然后被浸没辊浸入浆液中,这个浸没辊可以上下升降以允许纱线保持在浆液中所需时间[3]。然后,经纱片穿过两个被称为挤压辊的罗拉。挤压罗拉的目的是挤出多余的浆料,并用力将浆液挤压进纱线内部,使其完全渗透浆液[4]。烘房决定了最大生产率。要求快速、彻底、均匀地烘干潮湿的浆纱[5]。一个简单的双烘筒速度太慢,因为很难获得足够高的传热速率[6]。通过引入更多的烘筒,可以获得更多的干燥表面,并增加给定纱线速度的接触时间[7]。为了防止纱线之间的黏合,有必要将每根上过浆的经纱相互分隔开,然后才能将经纱用于织造,这是由分绞棒完成的[8]。

【Key Words】

ingredient [inˈgriːdiənt] 原料,成分
adhesive [ədˈhiːsiv] 黏合剂
lubricant [ˈluːbrik(ə)nt] 润滑剂
additive [ˈædətiv] 助剂
solvent [ˈsɔlvənt] 溶剂
slasher [ˈslæʃə(r)] 浆纱机
beam creel 经轴架
size box 浆槽
drying section 烘房
yarn separation section 纱线分绞区
headstock 机头
size liquor 浆液

dip roll = immersion roll 浸渍辊,浸没辊
moved up or down 上下升降
warp sheet 经纱片
squeeze [skwiːz] roll 挤压罗拉,挤压辊
physically [ˈfizikli] 物理上(用力的)
penentration [ˌpeniˈtreiʃn] 渗透,穿透
throughput [ˈθruːput] rate 生产率
rapidly 快速地
thoroughly [ˈθʌrəli] 彻底地,完全地
uniformly [ˈjuːnifɔːmli] 一致地,相同地
2-cylinder 双烘筒
available [əˈveiləbl] 可获得的,可利用的

【Key Sentences】

1. In general, size **ingredients** can be divided into four categories: **adhesives, lubricants, additives, and solvent.**

Fig 4.12 shows the different sizes.

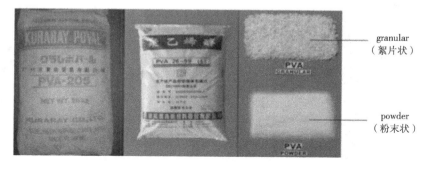

Fig 4.12 Size(浆料)

2. The beam creel is merely a device of frame on which warp beams are placed in a manner convenient for unwinding.

Fig 4.13 shows the sizing workshop.

Fig 4.13　Sizing workshop(浆纱车间)

3. The yarn is fed into the size box by means of a guide roll. It then passes under a **dip** or **immersion roll** which is capable of being **moved up and down** allowing the yarn to be held in the size liquor for a desired period of time.

Fig 4.14 shows the princple of sizing. Fig 4.15 shows the princple of size imprgnation.

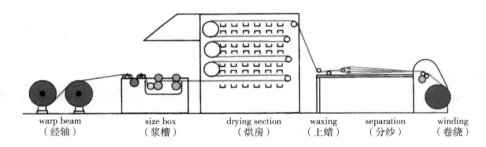

Fig 4.14　Princple of sizing machine(浆纱机工作原理)

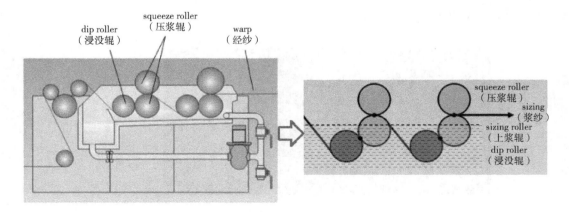

Fig 4.15　Princple of size imprgnation(浆料浸渍原理)

4. The purpose of the squeeze rolls is to squeeze out of **excess** size and to **physically** drive the size into the yarn for proper **penentration**.

5. The drying section determines the maximum **throughput rate**. It is required to dry the wet sized yarn **rapidly**, **thoroughly** and **uniformly**.

6. A simple 2-**cylinder** machine is too slow as it is difficult to get a sufficiently high heat transfer rate.

7. By introducing more cylinders, more drying surface is made **available** and the contact time for given yarn speed is increased.

8. In order to prevent adhesion between the yarns, it is necessary to separate each sized end from the others before the warp can be used for weaving, which is performed by the **burst rod**s.

Part 2 *Trying*

1. Please sort the sizing requirements of the following different kind of yarn from high to low? Why are you arranging this?

	28 tex carded yarn	18 tex combed yarn	18 tex compact yarn	26 tex twistless yarn
sort				
reason				

2. How many kinds of size are on the market at present? Find out 3~4 species, which fiber are suitable for respectively?

	Specific name	Fiber (suitable for this size)
Size one		
Size two		
Size three		
Size four		

3. Under the guidance of the training teacher, learn the operation of sizing prototype(小样机) in the training factory.

Part 3 *Thinking*

1. According to the knowledge you learned in this project, what advantages the warp sizing bring to the following process from the economic benefits, especially the weaving process?

_____
_____
_____

2. Please find some brand of sizing machine from website and talk about their structure with your classmates.

_____
_____
_____

Task Three Woven Technology(织造技术)

【Text】

机织物是在织布机上完成生产的,虽然织布机以多种方式进行过改变,但是最基本的原理和操作至今依然保持不变[1]。织轴上的经纱绕过后梁,向前穿过经纱自停装置上的停经片以及综丝,综丝主要负责分离经纱片,目的是形成开口[2]。然后,穿过钢筘,它保持纱线之间均匀的间距,并负责对已经远离由两层经纱片和钢筘形成的三角形经纱开口区的纬纱打纬[3]。边撑在织口稳定地握持布边,以帮助形成均匀的织物,然后经过胸梁,绕过卷曲辊卷绕在卷布辊上[4]。边撑在织口处紧紧地握持住已经形成的织物,以利于所形成织物具有很好的均匀性。

【Key Words】

loom [luːm] 织布
warp beam 织轴
back rest 后梁
come forward 向前
drop wire 停经片
heald [hild] 综丝
shed formation 开口成形

reed 钢筘
beating-up 打纬
temples ['templz] 边撑
fell 织口
front rest 胸梁
take-up roller 卷布辊
cloth roller 布辊

【Key Sentences】

1. Weaving is done on a machine called a **loom**, which has changed in many ways, but remains the basic principles and operations.

2. The yarn from the **warp beam** passes round the **back rest** and **comes forward** through the

drop wires of the warp stop-motion to the **healds**, which are responsible for separating the warp sheet for the purpose of **shed formation**.

Fig 4.16 shows the diagram of all loom parts.

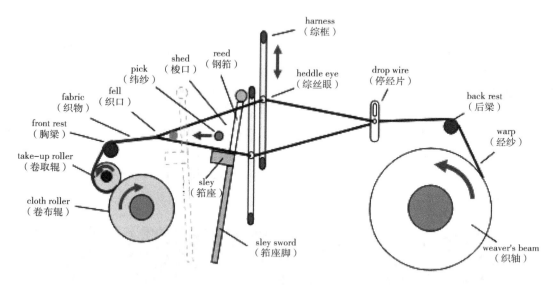

Fig 4.16　Diagram of all loom parts(织机所有部件简图)

3. It then passes through the **reed**, which holds the threads at uniform spacing and is responsible for **beating-up** the weft thread that has been left in the triangular warp shed formed by the two warp sheets and the reed.

4. **Temples** hold the cloth firmly at the **fell** to assist in the formation of a uniform fabric, which then passes over the **front rest**, round the **take-up roller**, and onto the cloth roller.

【Text】

为了在任何型号的织机上使经纱交织而生产织物,下面三个操作是很重要的。开口:开口运动将经纱分成两层(上层和下层),形成梭口[1];引纬:引纬是一种牵引纬纱穿过梭口、横穿织物的运动[2];打纬:打纬推动新穿入的纬纱,将其打入织物已形成的织口[3]。这三个操作工序通常被称为织造的基本运动,且必须按一定顺序进行,但是它们彼此之间的精确时机控制是非常重要的,并且要详加考虑[4]。

【Key Words】

interlace [ˌintəˈleis] 使交错,使交织
shedding [ˈʃediŋ] 开口
shed　(织机的)梭口,梭道
picking-filling insertion　引纬
travers across　横越,横穿
beating-up　打纬

insert [inˈsə:t] 插入,传入
pick [pik] 纬纱
fell [fel] 织口
primary [ˈpraiməri] 首要的,主要的
sequence [ˈsi:kwəns] 顺序,序列
precise [priˈsais] 精密的,精确的

extreme [ikˈstriːm] 极端的,过激的,极限的,非常的
additional [əˈdɪʃənl] 补充, 额外的,附加的
essential [ɪˈsenʃl] 必要的, 基本的

【Key Sentences】

1. The **shedding** motion separates the warp threads into two layers (an upper and a lower layer) to form a **shed**.

2. **Picking** is the operation of passing the weft, which **traverses across** the fabric, through the shed.

3. **Beating-up** pushes the newly **inserted** weft, known as the **pick**, into the already woven fabric at a point known as the **fell**.

Fig 4.17 shows the three-dimensional diagram of weaving. Fig 4.18 shows the amplification diagram of shed.

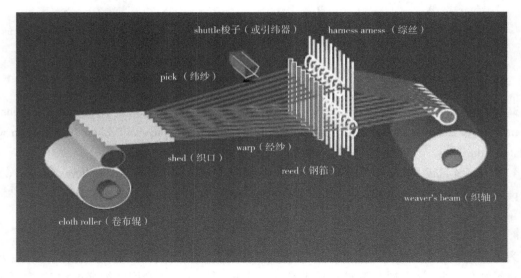

Fig 4.17　Three-dimensional diagram of weaving（织造三维图）

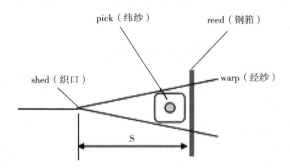

Fig 4.18　Amplification diagram of shed（梭口放大图）

4. These three operations are often called the **primary** motions of weaving and must occur in a given **sequence**, but their **precise** timing in relation to one another is also of **extreme** importance and will be considered in detail.

【Text】

如果织造要连续进行,还有两个必不可少的额外步骤[1]。送经:这个运动将经纱送到织造区域并以一个适当的、持续的张力在带边织轴上将其退绕[2]。卷取:这个运动以恒定速度从织造区域卷取织物,可以提供一个需要的纬纱间距,然后将织物卷到卷布辊上[3]。

【Key Words】

additional [ə'diʃənl] 附件的,额外的
essential [i'senʃl] 完全必要的,必不可少的
warp control = let-off 送经
unwind [ˌʌn'waind] 退绕
flanged [flændʒd] 带边缘的

weaver's beam 织轴
cloth control = take-up 卷取
withdraw [wið'drɔː] 撤走,拿走
constant ['kɔnstənt] 永恒的,始终如一的
pick-spacing 纬纱间距
roller 卷布辊

【Key Sentences】

1. Two **additional** operations are **essential** if weaving is to be continuous.

2. **Warp control** (or **let-off**) delivers warp to the weaving area at the required rate and at a suitable constant tension by **unwinding** it from a **flanged** tube known as the **weaver's beam**.

3. **Cloth control** (or **take-up**) **withdraws** fabric from the weaving area at the **constant** rate that will give the required **pick-spacing** and then winds it onto a **roller**.

【Text】

引纬是纬纱穿过打开梭口的过程[1]。传统的引纬方式是通过梭子,织机的新发展集中围绕通过其他载纬器取代梭子[2]。将不用梭子作为载纬器的织机根据无梭织机的引纬头进行分类[3]。如今,一般的引纬方法包括传统的梭子、剑杆、片梭、喷气和喷水,后四种属于无梭系统[4]。

织机设计者一直思考以纬纱引入的方式来取代梭子[5]。主要原因在于生产效率的提高。物理定律指明,要想穿过梭口,梭子必须加速和减速[6]。这自然需要花费时间与能量。

【Key Words】

center around 集中于⋯,包围,围绕
be grouped 分组,分类
conventional [kən'venʃənl] 传统的,习惯的,常规的
rapier ['reipiə(r)] 剑杆
gripper ['gripə] 夹具

projectiles ['prɔdʒiktailz] 投射物
gripper projectiles 片梭
physical laws 物理学规律
accelerate [ək'seləreit] 加速加快
decelerate [ˌdiː'seləreit] 减速,减慢

【Key Sentences】

1. **Filling insertion** is the passing of the filling yarn through the open shed.
2. While the traditional method of filling insertion is by means of a shuttle, new development in weaving have **centered around** replacing the shuttle with other filling carries.

Fig 4.19 shows the shutttle on traditional weaving machine.

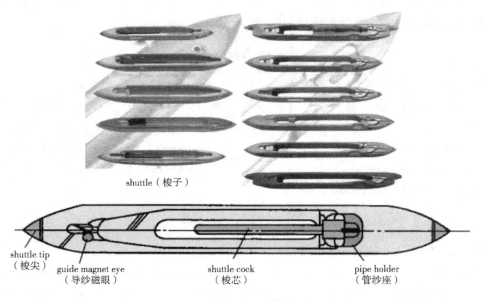

Fig 4.19 Shuttle(梭子)

3. Looms using a filling carrier other than a shuttle are **group**ed under the general heading of shuttless looms.
4. The common filling insertion methods in use today include the **conventional** shuttle, **rapiers**, **gripper projectiles**, air jet and water jet, the last four being shuttleless systems.

Fig 4.20 shows the gripper projectile. Fig 4.21 shows the rapier.

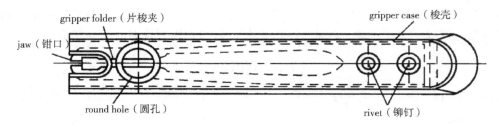

Fig 4.20 Gripper projectile(片梭)

5. Loom designers have constantly sought to replace the shuttle as a means of filling insertion. Fig 4.22 shows the process sketch of gripper insertion.

Project Four Woven Fabrics and Woven Technology(机织物与机织技术)

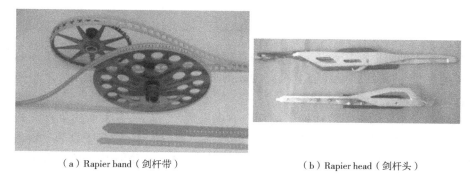

(a) Rapier band (剑杆带)　　　　(b) Rapier head (剑杆头)

Fig 4.21　Rapier(剑杆)

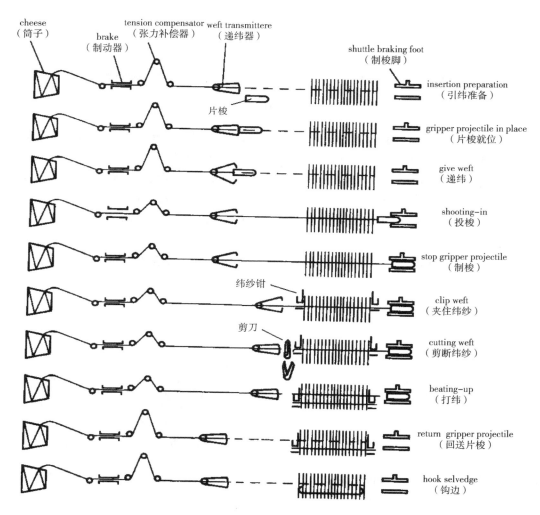

Fig 4.22　Process sketch of gripper insertion(片梭织机引纬过程简图)

6. **Physical laws** dictate that, to move through the shed, the shuttle must be **accelerated** and **decelerated.**

137

【Text】

如果只花很少的时间对梭子进行加速,并且梭子穿过梭口的时间可以减少,那么可以获得较高的织机速度[1]。随着纬纱载纬器上供应卷装的去除,且载纬器可以制作得更小些以至于减少纱线在梭口的运动[2]。

无梭织机技术正在向企图取代梭子的两个方向发展[3]。首先,在发明的剑杆织机和片梭织机中,梭子被另外一个坚固的载纬器所取代[4]。在流体喷射织机中,较高压力的气流或水流被用于携带纬纱穿过梭口,这是第二个方向[5]。流体喷射织机允许单位时间内较高数量的纬纱但对纱线类型有一定限制[6]。

【Key Words】

transit ['trænzit] 运输,通过,经过
filling carrier 纬纱载纬器
proceed [prə'siːd] 行进,前往

rapier ['reipiə(r)] loom 剑杆织机
gripper ['gripə] loom 片梭织机
fluid jet loom 流体喷射织机

【Key Sentences】

1. If less time could be spent accelerating the shuttle, and the **transit** time of the shuttle through the shed could be reduced, higher loom speeds could be obtained.

2. With the removal of the filling supply package from the **filling carrier**, the filling carrier could be made smaller so the yarn movement in shedding could be reduced.

3. Shuttleless loom technology has **proceeded** in two general directions in its attempt to relpace the shuttle.

4. In the first, as found in **rapier and gripper looms**, the shuttle is replaced by another solid carrier.

Fig 4.23 shows the schematic of rapier insertion.

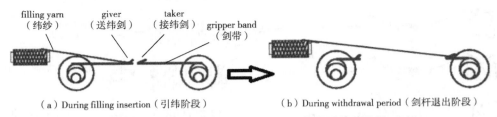

(a) During filling insertion (引纬阶段)　　(b) During withdrawal period (剑杆退出阶段)

Fig 4.23 Schematic of rapier insertion (剑杆引纬示意图)

5. **Fluid jet looms**, in which a high pressure jet of air or water is used to carry the filling through the shed, represent the second direction.

Fig 4.24 shows the schematic of air jet insertion or water jet insertion. Fig 4.25 shows the work diagram of air jet insertion.

6. Fluid jet looms allow high numbers of picks per unit time but have some limitations as to yarn type.

Project Four Woven Fabrics and Woven Technology(机织物与机织技术)

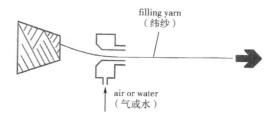

Fig 4.24 Schematic of air jet insertion or water jet insertion(喷气或喷水引纬示意图)

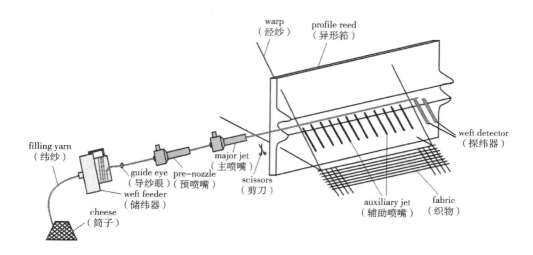

Fig 4.25 Work diagram of air jet insertion(喷气引纬工作简图)

【Text】

在喷气织机中,高压气流携带纬纱穿过梭口。这种类型的织机的主要缺陷是气流扩散迅速以至于织物宽度受限[1]。现代喷气织机使用气流引导器或气流引导器加辅助喷嘴来延伸距离,这样纱线可携带[2]。

在喷水织机中,纬纱是被高压的水流所携带。由于水的扩散作用较气流慢,因此,喷水织机不要引导器可加工更宽的织物[3]。由于水是载纬器,因此,对纱线的种类以及纱线织前工序,如浆纱,有限制[4]。

【Key Words】

drawback ['drɔːbæk] 缺点
diffuse [di'fjuːz] 扩散,分散
auxiliary jet 辅助喷嘴

restriction [ri'strikʃn] 限制,约束
pre-weaving 织前

【Key Sentences】

1. The major **drawback** to air jet loom is that the air **diffuses** quite rapidly so the width of the fabric is limited.

Fig 4.26 shows the air jet loom.

Fig 4.26 Air jet loom(喷气织机)

2. Modern air jet looms use air guides or air guides with **auxiliary jets** to extend the distance that the yarn can be carried.

Fig 4.27 shows the major jet and auxiliary jet.

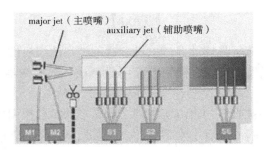

Fig 4.27 Major jet and auxiliary jet(主喷嘴与辅助喷嘴)

3. Since water diffuses much more slowly than air, wider fabrics can be made with water jet looms without using guides.

Fig 4.28 shows the water jet loom.

4. However, because water is the carrier, there are **restrictions** as to yarn types and **pre-weaving** yarn processing, such as slashing.

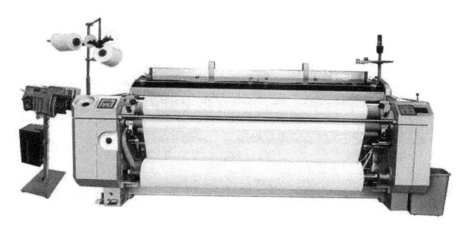

Fig 4.28 Water jet loom(喷水织机)

Part 2 *Trying*

1. Please use the flowchart to mark what parts the warp yarn needs to go through in the weaving process?

___

___

___

___

2. According to the knowledge you learned in this project and through the online search of relevant information, list the fiber and yarn varieties suitable for processing of rapier loom, air jet loom and water jet loom. And briefly explain the reason?

Types of looms	The fiber and yarn varieties suitable for processing	Reasons
Rapier loom		
Air jet loom		
Water jet loom		

3. Operate the fabric prototype(小样机) of the training factory and pay attention to the five major movements of weaving.

Part 3 *Thinking*

1. From Economic perspectives, what opportunities do the shuttless looms such as air jet looms and water jet looms bring to the weaving enterprises in the transformation and upgrading?

_____
_____
_____

2. Concatenate the knowledge of the previous projects, taking the rapier loom as an example, write down the weaving process of pure cotton woven fabric.

_____
_____
_____
_____

Task Four Woven Fabrics(机织物)

【Text】

传统的机织物是在织机上通过两组纱线以直角并按确定的顺序相互交织而成的织物结构[1]。纵向纱线被称作经纱,经纱中的一根可以叫作"an end"[2]。经纱是一组固定在织轴上的平行纱片。横向纱线被称作纬纱,一根纬纱可以叫作"a pick"[3]。通常每次每根纬纱与经纱成直角置于两组经纱片之间[4]。机织物的中心部分称为布身,边缘部分称为布边[5]。布边通常在纱线与织物结构上与布身都不同。广义上的机织物还有经纱不平行于布边,即经纱和纬纱互成一定的角度[6]。

机织物组织是表示经纱和纬纱的交织方式[7]。组织循环是最小的组织单元,当组织循环不断重复,就在面料上形成了设计所要求的图案[8]。有许多方式来表示织造,最常用的表示织造的方式是采用意匠纸[9]。在意匠纸上,正方形的垂直列表示经纱,而正方形的水平行表示纬纱[10]。通常使用填充的正方形表示经纱在纬纱之上,用空白正方形表示纬纱在经纱之上[11]。三种基本组织为平纹、斜纹和缎纹。

【Key Words】

repeat [ri'pi:t] 组织循环
interlace [ˌintə'leis] 交织,交错
longitudinal [ˌlɔŋgi'tju:dinl] 纵向的
an end 一根经纱
widthwise 与宽同方向地,纬向地

a pick 纬纱
in the broad sense of… 从广义上讲
uneven 不平行的
body 布身
selvages 布边

fabric weave = fabric construction 织物组织
manner 方式,方法
repeat 组织循环
represent [ˌrepriˈzent] 代表,表示
design paper 艺匠纸
vertical [ˈvəːtikl] column 垂直列
square [skweə(r)] 正方形

horizontal [ˌhɔriˈzɔntl] row 水平行
filled-in 已填满的
blank 空白的
basic weave 基本组织
plain weave 平纹
twill weave 斜纹
satin weave 缎纹

【Key Sentences】

1. The conventional woven fabric is a textile structure formed on a loom when two sets of yarns are **interlaced** at right angles in **established** sequences.

Fig 4.29 shows the warp and weft are interlaced at right angles.

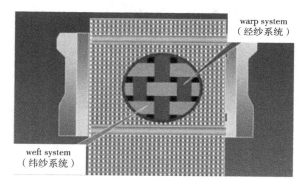

Fig 4.29 The warp and weft are interlaced at right angles(经纱与纬纱相互垂直)

2. The **longitudinal** yarns are known as the warp and one yarn of the warp is called **an end**.

3. The **widthwise** yarns are known as the filling or weft and one yarn of filling is known as a **pick**.

4. The filling yarns are laid between two warp sheets, usually one pick at a time, at right angles to the warp.

5. The central portion of the woven fabric is called the **body** and the edges are called the **selvages**.

6. **In the broad sense of** the woven fabric, the warp yarn is **uneven** with the selvages, so warp and weft are at a certain angle.

Fig 4.30 shows the warp and weft are a certain angle(30°).

7. The **fabric weave** is the **manner** in which the warp and weft yarns are interlaced.

Fig 4.31 shows the diagram of fabric weave.

8. The **repeat** is the smallest unit of the weave which will produce the design required in the fabric when repeated.

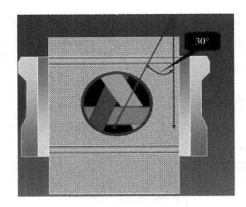

Fig 4.30 Warp and weft are a certain angle(30°)[经纱与纬纱形成一定角度(30度)]

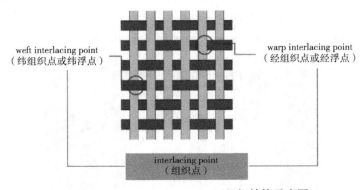

Fig 4.31 Diagram of fabric weave(组织结构示意图)

9. There are many ways of **representing** a weave. The most commonly used method of representing a weave is on **design paper**.

Fig 4.32 shows the design paper.

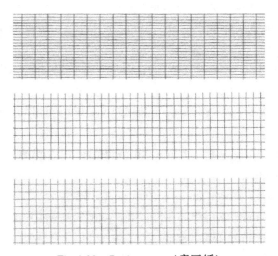

Fig 4.32 Design paper(意匠纸)

Project Four Woven Fabrics and Woven Technology(机织物与机织技术) ◀

10. On the design paper, the **vertical columns** of **squares** represent warp ends and the **horizontal rows** of squares represent filling picks.

11. It is normal to use a **filled-in** square to indicate that the end is over the pick and a **blank** square to indicate that the pick is over the end.

Fig 4.33 shows the three **basic weaves**(**plain weave**, **twill weave** and **satin weave**).

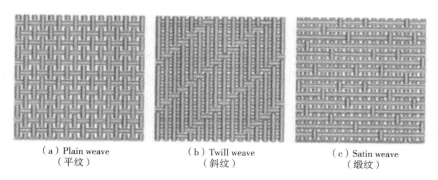

(a) Plain weave　　　　(b) Twill weave　　　　(c) Satin weave
　　(平纹)　　　　　　　　(斜纹)　　　　　　　　(缎纹)

Fig 4.33　The three basic weaves(三原组织)

【Text】

平纹是纱线交织的最简单形式，它是由经纱交替开口，而每根纬纱在相邻两根经纱间一上一下穿过组成后而形成织物[1]。相邻的纬纱具有相反的交织规律，就是说此纬纱位于某一经纱上方，而这根经纱又位于前一纬纱的上面，此纬纱位于某一经纱的下方，而这根经纱位于前一纬纱的下方[2]。平纹的一个组织循环包括两根经纱和两根纬纱。尽管平纹很简单，平纹织物在所有织物中占有较多的比例[3]。与其他织物相比，平纹织物具有较多的交织次数，导致其成为最结实的织物[4]。

多达70%的机织物由平纹组织构成。除非表面印有颜色，平纹织物正反面结构相同[5]。织造平纹组织织物时，纱线可以稀疏或者紧密排列，每10厘米内的经纱根数和纬纱根数可以相同，也可以变化很大[6]。纱线直径和纱线类型可以变化，以生产出新颖的织物。每10厘米内经纱数和纬纱数大致一样时，纱线密度也叫织物密度是平衡的；而每10厘米内经纬纱根数不相同时，织物密度是不平衡的[7]。平纹织物的生产相对便宜。通过上色和后整理，可以在整理织物上制作出许多(图案)设计[8]。另外，一些耐用织物的生产往往采用平纹组织。

【Key Words】

consist of　包括；由……组成
alternate [ɔːˈtəːnət]　交替的
adjacent [əˈdʒeisnt]　相邻，邻近的，毗邻的
reverse [riˈvəːs]　(使)颠倒；掉换，交换
preceding [priˈsiːdiŋ]　在先的，在前的
simplicity [simˈplisəti]　简单，容易

firm [fəːm]　结实的，坚固的
up to　多达
construct [kənˈstrʌkt]　构成，修建，建造
reversible [riˈvəːsəbl]　正反面一样的
pack 排列　loosely [ˈluːsli]　松散的，蓬松的
tightly [ˈtaitli]　坚固地，牢固地

145

considerably [kən'sidərəbli] 非常，相当多地
approximately [ə'prɔksimətli] 近似地，大约
thread count = fabric count 织物经纬密度
differ ['difə(r)] 不同，相异
considerably [kən'sidərəbli] 相当，非常
comparatively [kəm'pærətivli] 比较上，相对地

【Key Sentences】

1. The plain weave is the simplest form of yarn interlacing and **consists of** the **alternate** shedding of warp yarns to provide a fabric in which each filling yarn passes one warp and under the **adjacent** warp.

2. An adjacent filling yarn **reverses** the interlacing so that it passes over the warp that on the top of the **preceding** filling yarn and under the warp that lay under the preceding filling yarn.

3. Despite their **simplicity**, plain fabrics take a high percentage of all woven fabric.

4. Palin fabric has the highest number of interlacing as compared to other weaves causing it to be the **firmest**.

5. Unless colors have been printed onto the surface, plain weave fabrics are generally considered to be **reversible**.

Fig 4.34 shows the plain weave fabrics.

Fig 4.34 Plain weave fabric(平纹织物)

6. In making plain weave fabrics, yarns can be **packed loosely** or **tightly** together, and the number of warp yarns may equal to the number of filling yarns in each 10cm or may vary **considerably**.

7. When the number of warp yarns per 10cm is **approximately** the same as the number of filling yarns per 10cm, the **thread count**, sometimes called **fabric count**, is considered balanced; when the number of yarns per 10cm **differs considerably** between warp and filling, the fabric count is unbalanced.

8. A variety of designs can be made using various coloring and finishing processes on the finished fabric.

Project Four　Woven Fabrics and Woven Technology(机织物与机织技术)

【Text】

斜纹是第二种基本组织,斜纹的特征是在面料上有多条带有一定角度的对角线[1]。斜纹组织表示方法中的数字代表了第一根经纱遍及织造中所有纬纱的位置[2]。对简单的斜纹组织来说,一个循环由上下数字之和的经纱和纬纱数所组成[3]。例如,二上一下斜纹由3根经纱和3根纬纱反复循环。剩余经纱的位置是通过重复第一根经纱的运动所决定的,但是紧接着的纬纱开始运动[4]。由于每三根经纱中每一根织造都不一样,上面提到的织物至少需要三个综框来制作[5]。

【Key Words】

diagonal line　对角线
denote [di'nəut] 标志,象征,表示
represent [ˌrepri'zent] 代表
throughout [θru:'aut] 遍及,贯穿
remaining [ri'meiniŋ] 其余的
successive [sək'sesiv] 连续的,接连的,相继的

harness ['hɑːnːs] 综框
clue [kluː] 提示,暗示,线索
predominate [pri'dɔmineit]（数量上）占优势;以…为主
warp-faced twill　经面斜纹
filling-faced twill　纬面斜纹

【Key Sentences】

1. Twill is the second basic weave and is characterized by **diagonal lines** running at angles to the fabric. Fig 4.35 shows the twill weave fabric.

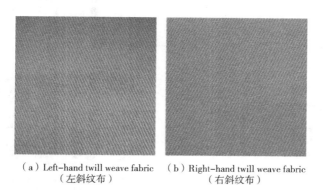

(a) Left-hand twill weave fabric
　　（左斜纹布）
(b) Right-hand twill weave fabric
　　（右斜纹布）

Fig 4.35　Twill weave fabric(斜纹织物)

2. These numbers **represent** the position of the first end **throughout** the number of picks in the weave.

3. For simple twill weaves, the **repeat** consists of as many ends and picks as the sum of the numbers in twill weave expestion.

4. The positions of the **remaining** ends can then be determined by repeating the motion of the first end but beginning the motion on **successive** picks.

5. Since each of the three ends weaves differently, the above fabric requires a minimum of three **harnesses** to produce. Fig 4.36 shows the warp-faced twill. Fig 4.37 shows the filling-facd twill. Fig 4.38 shows the even twill.

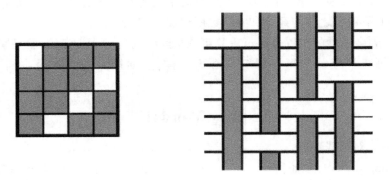

Fig 4.36　Warp-faced twill(经面斜纹)

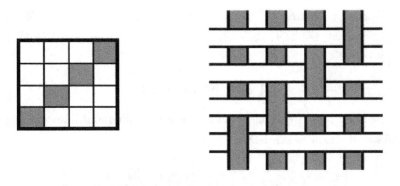

Fig 4.37　Filling-facd twill(纬面斜纹)

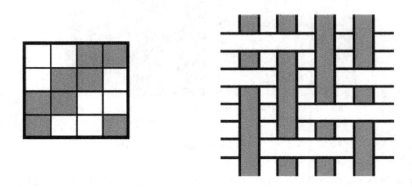

Fig 4.38　Even twill(双面斜纹)

【Text】

斜纹组织分左斜纹和右斜纹。如果对角线从织物的左下移动至织物的右上,则是右斜纹;

如果斜纹线从左上移至右下则是左斜纹[1]。双面斜纹织物中,处于纬纱上面和下面的经纱根数是相同的,而在单面斜纹织物中,从经纱上面穿过的纬纱多(或少)于从经纱下面穿过的纬纱[2]。单面斜纹组织有正反面之分,所以正反面结构不同[3]。纬面斜纹中,纬纱在织物正面占优势,经面斜纹的正面经纱较多[4]。双面斜纹正反面经纬纱根数相同,就认为正反面结构相同。

斜纹线可以从15°的缓斜纹变化到75°的急斜纹[5]。最常见的45°角的斜纹称为正则斜纹[6]。斜线的角度取决于经纱的紧密程度,每10cm内所用纱线根数,所用纱线的直径和形成(组织)循环的实际情况[7]。斜纹织物有独特的吸引人的外观[8]。总的来说,斜纹交织织物结实耐用。斜纹与平纹的不同之处在于完成一个组织循环图案所需纬纱根数和经纱根数[9]。

用于斜纹组织织物的纱线通常高捻高强[10]。由于交织少,斜纹组织允许纱线排列更紧密。这些高强纱紧密地排列形成结实耐用的织物。但是,如果排列过紧,织物撕破强力、耐磨性能和褶皱回复性会下降[11]。与平纹织物相比,斜纹织物除了良好的外观和较高的强度之外,它还不容易被沾污[12]。但由于需要更复杂的织造技术和织机,斜纹织物生产成本比平纹高[13]。

【Key Words】

right-hand twill　右斜纹
left-hand twill　左斜纹
lower left　左下方
upper right　右上方
upper left　左上方
lower right　右下方
even twill　双面斜纹
uneven twill　单面斜纹
more or fewer　或多或少
a right and a wrong side　正反面
reversible [ri'və:səbl]　正反面一样的
filling-faced twill　纬面斜纹
predominate [pri'dɔmineit]　(数量上)占优势

warp-faced twill　经面斜纹
reclining twill　缓斜纹
steep twill　急斜纹
regular twill　正则斜纹
closeness ['kləusnəs]　紧密程度
progression [prə'greʃn]　发展,进展
distinctive [di'stiŋktiv]　独特,与众不同的
attractive [ə'træktiv]　吸引人的
tearing strength　撕裂强度
abrasion resistance　耐磨性
wrinkle recovery　褶皱回复性
show soil　玷污,污染

【Key Sentences】

1. If the diagonal moves from the **lower left** to the **upper right** of the fabric, it is referred to as a right-hand twill; if it moves from **upper left** to **lower right**, it is a left-hand twill.

Fig 4.39 shows the left-hand and right-hand twill weave.

2. In **even twill** fabrics, the filling yarns pass over and under the warp yarns of the same number, whereas in **uneven twill** fabrics, the pick goes over **either more or fewer** warps than it goes under.

3. **Uneven twill fabrics** have **a right and a wrong side** and therefore are not considered **reversible**.

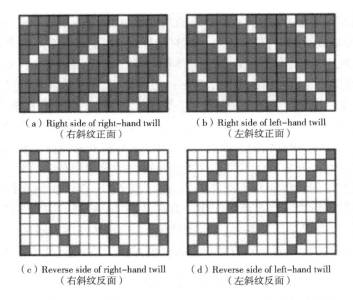

Fig 4.39 Left-hand and right-hand twill weave(左斜纹与右斜纹)

4. **Filling-faced twills** are those in which the picks **predominate** on the face of the fabric, and **warp-faced twills** have a more evident warp on the surface.

5. The diagonal line of the twill can vary from **reclining twill**, with a low 15-degree angle; to **steep twill**, with a 75-degree angle.

6. A twill angle of 45- degree, which is the most common, is considered to be a **regular twill**.

7. The angle of the diagonal is determined by the **closeness** of the warp ends, the number of yarns used per 10cm, the diameter of the yarns used, and the actual **progression** forming the repeat.

8. Twill weave fabrics have a **distinctive** and **attractive** appearance.

9. Twills differ from plain weave in the number of filling picks and warp ends needed to complete a repeat pattern.

10. Yarns used in twill weave fabrics frequently are tightly twisted and exhibit good strength.

11. However, if the yarns are packed too tightly, the fabric will have reduced **tearing strength**, **abrasion resistance** and **wrinkle recovery**.

12. In addition to appearance and good strength, twills tend to **show soil** less quickly than plain weave fabrics.

13. However, twills are more expensive to produce than plain weave fabrics, because of the more complicated weaving techniques and looms are required.

【Text】

缎纹组织构成了第三大类基本组织。缎纹组织的特征是具有较长的经浮长线和纬浮长线，并且交织点不连续[1]。这减少了对角线效应在织物正面发生的可能。但是，一些缎纹确实有模糊的对角线(斜纹)效应，因为少量纱线用于形成组织循环[2]。

缎纹可以分为经面缎纹与纬面缎纹[3]。经面缎纹中，经纱浮在表面。缎纹组织上，纬纱浮在织物表面的变化指的是纬面缎纹[4]。长丝纱常用于缎纹织物,而短纤纱,如棉,常见于纬面缎纹组织织物。但也有例外,棉纤维缎纹和长丝纬缎织物已被生产出来[5]。

【Key Words】

be charaterized by　具有……的特征
floats [fləʊts] 浮长线
nonadjacent [ˈnɒnəˈdʒeisənt] 不邻近的,不毗连的
faint [feint] 不清楚的

owing to　因为;由于
warp-faced satin　经面缎纹
weft-faced satin=sateen　纬面缎纹
variation [ˌveəriˈeiʃn] 变化,变更,变异

【Key Sentences】

1. Satins **are charaterized by** having **long floats** of either warp or filling yarn and **nonadjacent** interlacing points.

Fig 4.40 shows one repeat of a single satin weave.

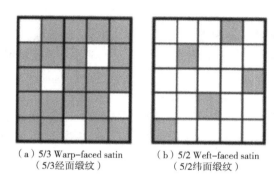

(a) 5/3 Warp-faced satin
（5/3经面缎纹）

(b) 5/2 Weft-faced satin
（5/2纬面缎纹）

Fig 4.40　One repeat of a single satin weave(单个缎纹组织循环)

2. However, some satins do have a **faint** diagonal effect **owing to** the low number of yarns used to form the repeat.

3. Satins can be divided into **warp-faced satin** and **weft-faced satin**, which is called **sateen**. Fig 4.41 shows the warp face satin and weft face satin.

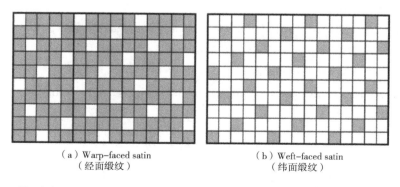

(a) Warp-faced satin
（经面缎纹）

(b) Weft-faced satin
（纬面缎纹）

Fig 4.41　Warp-faced satin and weft-faced satin(经面缎纹与纬面缎纹)

4. A **variation** of the satin weave in which the filling picks float on the surface is referred to as sateen.

5. There are, however, exceptions, cotton satins as well as filament yarn sateens have been made.

【Text】

缎纹组织中的浮线会产生一个有光泽的且易反光的表面[1]。如果浮线是低捻度（20~40捻/米）且明亮的长丝纱，这一特性会得到加强[2]。光泽使得织物更加适合做时髦服装，它的光滑性可用于做衬里。经缎和纬缎织物中的浮线容易引起钩丝、磨损[3]。浮长线越长，织物表面钩丝、变粗糙、显示磨损迹象的机会就越大[4]。因此，这些织物没有平纹织物和斜纹织物耐磨。

缎纹织物经常被用作外套、西装和夹克的衬里面料，因为缎纹可以很容易地使这些服装在其他材料上滑来滑去[5]。经面缎纹或纬面缎纹有一个明确的正面和反面。缎纹织物的变化可以通过使用高捻纬纱来制作，以形成绉缎织物[6]。

【Key Words】

shiny ['ʃaini] 光的,光亮的；闪耀的
accentuate [ək'sentʃueit] 强调,着重指出
reflect [ri'flekt] 反射
dressy ['dresi] 漂亮精致的,穿着时髦的
dressy wear 时髦服装
lining ['lainiŋ] 衬里
snag ['snæg] 被钩丝,被钩住
roughen 变粗糙；使粗糙
show signs of wear 显示磨损迹象
abrade [ə'breid] 磨损
slip [slip] 滑动,滑离
crepe-back satin 绉缎

【Key Sentences】

1. The long floats of the satin weave create a **shiny** surface and tend to **reflect** light easily.

2. This is **accentuated** if bright filament yarns with a low twist (about 20~40 turns per meter) are used for the floating yarns.

3. Floats in satin and sateen fabrics tend to **snag** and **abrade** easily.

4. The longer the float, the greater the chances that the surface of the fabric will **snag**, **roughen**, and **show signs of wear**.

5. Satins are frequently used as lining fabrics in coats, suits and jackets, as they make it easy to **slip** the apparel item on and off over other materials.

Fig 4.42 shows the satin weave fabric.

6. Variations in satin fabrics can be produced by using highly twisted yarns in the filling to create a **crepe-back satin**.

Part 2 *Trying*

1. Find the relevant material, point out the fabric weave used in the following fabrics and fill in the corresponding box. (卡其 khaki、牛仔布 denim、哔叽 serge、华达呢 gabardine、啥味呢 worsted

Project Four　Woven Fabrics and Woven Technology(机织物与机织技术)

Fig 4.42　Satin weave fabric(缎纹织物)

flannel、制服呢 uniform cloth、府绸 poplin、凡立丁 valtine、派立司 palace、电力纺 habotai、乔其纱 georgette、塔夫绸 taffeta、双绉 crepe de chinf、横贡缎 sateen、直贡缎 satin drill、素绉缎 plain satin)

Plain weave	Twill weave	Satin weave

2. Go to the market and purchase 3 to 4 kinds of three basic weave fabrics, observe their structure by the fabric density mirror and draw the weave structure of the fabrics.

Sample 1	Fabric weave	Sample 2	Fabric weave
Sample 3	Fabric weave	Sample 4	Fabric weave

3. Pleaseuse the fabric prototype to weave three basic weave fabrics.

Part 3 *Thinking*

1. According to the knowledge learned in the textbook, assuming that other factors are the constant, compare the three basic weaves fabrics from the fracture strength, tear strength, wrinkle resistance, luster and snagging.

_____
_____
_____
_____

2. List the clothing fabrics you come into contact with in your daily life (such as shirts, suits, jeans.) and analyze which kind of basic weave they might use based on the wearing performance. Why?

_____
_____
_____
_____

Project Five　Knitted Fabrics and Knitting Technology(针织物与针织技术)

Task One　Knitting Fabrics(针织物)

【Text】

针织是第二种广泛使用构成织物的方法。针织由纱线线圈互相串套而形成织物[1]。针织（用）纱线类型包括扁平长丝、变形长丝、短纤纱、天然纤维和再生纤维的混纺纱，以及新开发的纤维，如吸湿快干纤维(Coolmax)、PBT、PTT、莫代尔、抗菌纤维(Amicor)、竹纤维和大豆纤维等[2]。此外，现有的针织机已能成功编织各种高性能纤维，包括玻璃纤维、碳纤维、芳纶甚至陶瓷纤维[3]。

针织物的用途包括袜子、内衣、毛衣、长裤、西装和外套，以及地毯和其他家居用品[4]。此外，针织物在工程结构上的应用很有潜力。而且可以预见，针织品将继续流行，并在未来几年的各种应用中被发现[5]。

【Key Words】

be defined to　被定义为
intermesh [ˌɪntəˈmeʃ] 使互相啮合，此处可译为串套
flat filament　扁平长丝
textured filament　变形长丝
coolmax　一种吸湿快干纤维
amicor　一种抗菌纤维
soybean fiber　大豆纤维
aramid [ˈɛrəmɪd] 芳纶
ceramic [səˈræmɪk] 陶瓷(纤维)
hosiery [ˈhəʊzɪəri] 袜子
underwear　内衣
slacks [slæks] 宽松的长裤
suits　西服
rugs [rʌgz] 地毯
furnishings　家具用品

【Key Sentences】

1. Knitting can be defined to be the formation of fabric by the **intermeshing** of loops of yarn.

2. Knitting yarn types include **flat and textured filament**, spun, blends of natural and man-made fibers, and new developed fibers such as **Coolmax**, PBT, PTT, Modal, **Amicor**, bamboo fiber and **soybean fiber**, etc.

3. Furthermore, existing knitting machines have been successfully adapted to use various types of high-performance fibers, including glass, carbon, **aramid** and even **ceramic**.

4. The usage of knitted fabrics ranges from **hosiery**, **underwear**, sweaters, **slacks**, **suits**, and coats, to **rugs** and other home **furnishings**.

Fig 5.1 shows all kinds of knitting products.

Fig 5.1 Knitting products(针织产品)

5. And it can be expected that knits will continue to be popular and to be found in a wide variety of applications in the years to come.

【Text】

针织线圈结构介绍如下:纬编针织线圈包括圈干和沉降弧,经编针织线圈包括圈干和延展线[1]。针织不像机织那样,需要两套纱线,针织只需要一套纱线即可[2]。这一套纱线可以包括一根纱线(纬编)或一组纱线(经编)[3]。在经编和纬编工艺中,用于形成线圈的主要机件是织针。在经编和纬编中最常用的织针是舌针[4]。

【Key Words】

element [ˈelimənt] 要素
knitted loop 针织线圈
needle loop 圈干
sinker loop 沉降弧
underlap loop 延展线
sets [ˈsets] 套

weft knitting 纬编
warp knitting 经编
principal [ˈprinsəpl] 主要的
mechanical elements 机械零件
latch [lætʃ] needle 舌针

【Key Sentences】

1. As for a weft knitted loop, it consists of a **needle loop** and a **sinker loop**, while for a warp

Project Five Knitted Fabrics and Knitting Technology(针织物与针织技术)

knitted loop, it consists of a needle loop and a **underlap loop**.

Fig 5.2 shows the weft knitting loop. Fig 5.3 shows single weft knitting unit.

2. Unlike weaving, which requires two yarn **sets**, knitting is possible using only a single set of yarns.

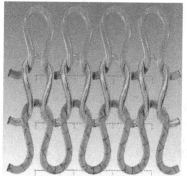

Fig 5.2   Weft knitting loop(纬编线圈)

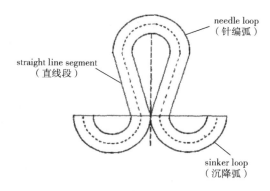

Fig 5.3   Single weft knitting unit(一个纬编线圈单元)

3. The set may consist of a single yarn (**weft knitting**) or a single group of yarns (**warp knitting**).

4. The most common type of needles, used in both warp and weft knitting, is the **latch needle**.

【Text】

针织物可以分为纬编织物和经编织物[1]。在纬编过程中,线圈由单根纬纱形成。形成的线圈,多少不限,沿着织物的宽度方向形成一排水平的线圈,即线圈横列[2]。也就是说,纬编织物中纱线是沿织物横向编织成圈,形成织物。线圈总是沿织物横向形成一排排水平排列的线圈。

在经编中,线圈沿着垂直方向或经向而成[3];所有的线圈同时形成一个线圈横列,同时线圈纵列的长度也在增加[4]。加工过程中,织物宽度所需的一组或多组纱线相互平行排列在经轴上,这个经轴类似于织造用的经轴[5]。经轴放置于编织区域的后方和上方,所有的纱线同时喂入到针织区域,最终将纱线编织成织物。每根纱线由一根特殊的织针操控,所有的纱线同时形成线圈[6]。

总之,纬编的特点是通过纬纱喂入形成的线圈与织物形成方向成直角[7]。相反,经编的特点是通过经轴上经纱喂入形成的线圈与织物形成方向平行[8]。更精确地说,在针织过程中,经编受将每根纱线交织成相邻纵行的影响[9]。

【Key Words】

classification [ˌklæsifiˈkeiʃn] 分类,分级,类别
weft-knitting stitch 纬编织物
warp-knitting stitch 经编织物
horizontal row [ˌhɔriˈzɔntl] 水平行
course [针]线圈横列
simultaneously [ˌsiməlˈteiniəsli] 同时
wale [针]线圈纵行
resemble [riˈzembl] 像
manipulate [məˈnipjuleit] 控制,操作

【Key Sentences】

1. Knitted fabrics can be classified as **weft-knitting stitch** and **warp-knitting stitch**.

2. The loops are formed, more or less, across the width of the fabric usually with **horizontal rows** of loops.

3. In warp knitting, the loops are formed in a vertical or **warpwise** direction.

4. All of the loops making up a single **course** are formed **simultaneously**, thus the lengths of each vertical column of loops, the **wales**, increase at the same time.

5. One or several groups of yarns required for the width of the fabric under construction are placed parallel to each other on a beam that **resembles** a warp beam for weaving.

6. Each yarn is **manipulated** by one specific needle, and all the yarns form loops at the same time.

7. Weft knitting is characterized by loops forming through the feeding of the weft yarn at right angles to the direction in which the fabric is produced.

Fig 5.4 shows the schematic of weft knitting fabric.

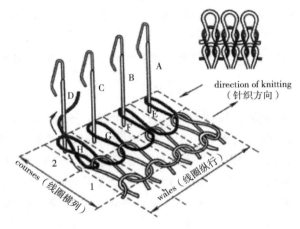

Fig 5.4 Schematic of weft knitting fabric(纬编织物的原理)

Project Five Knitted Fabrics and Knitting Technology(针织物与针织技术)

8. Warp knitting, on the other hand, is characterized by loops formed by the warp yarns from warp beams, parallel to the direction in which the fabric is produced.

Fig 5.5 shows the schematic of warp knitting fabric.

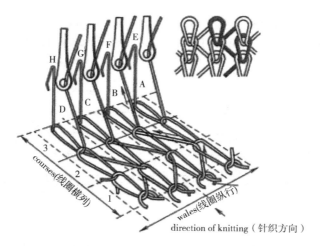

Fig 5.5 Schematic of warp knitting fabric(经编织物的原理)

9. More precisely, warp knitting is affected by interlooping each yarn into adjacent columns of wales during knitting progresses.

Part 2 *Trying*

1. In the training Center or laboratory, looking for a piece of woven fabric and a piece of knitted fabric, analyzes the difference of the structure between them in detail.

_____
_____
_____
_____

2. According to the structural characteristics of the knitted fabrics, go to the textile and garment market, then investigate and survey on knitting products, how many knitted products do you found in the market? What do you feel?

_____
_____
_____
_____

Part 3 *Thinking*

1. From the economic point of view, analyzes the cost of woven technology and knitting technology? which is higher? Why?

2. Go to the market and purchase the knitted fabrics and the woven fabrics with the same raw materials. What is the performance difference between the two fabrics by hand sensing and visual measurement?

Task Two Knitting Technology(针织技术)

【Text】

针织机件实际上包括织针、沉降片、固定织针的针床或骨架,以及处于针织区域放置纱线的储纱器[1]。在经编和纬编中,用于形成线圈的主要机件是织针。在现代针织中,三种主要的织针分别是舌针、钩针和复合针[2]。

纬编织物一般采用舌针。舌针形成线圈的运动过程如下:在起针位置,被握持的线圈停在开口针舌上面,在针向上运动时,被握持的线圈从针舌上滑到针杆上,为退圈[3]。针钩向下运动使其握持一段新的纱线,此过程称为喂纱[4]。当针继续向下运动时,在旧线圈的作用下,针舌被关闭[5]。当被握持的线圈与针分开时完成脱圈[6]。脱圈之后,线圈被拉伸,一个新的线圈形成,此时针必须返回到起针位置,这样就完成了一个循环[7]。

【Key Words】

sinker 沉降片	stem 针杆
latch needle 舌针	upwards 向上
spring-bearded needle 钩针	downwards 向下
compound needle 复合针	engage[inˈgeidʒ] 雇佣,聘用,这里可译文使用
running position 运行位置,起始位置	feeding [针]喂纱
rest on 搁在[支撑在]…上,依赖在…上	knockover [ˈnɔkəuvər] 脱圈
clearing [针]退圈	disengage [ˌdisinˈgeidʒ] 分开,释放
slip off 滑脱	pulling 拉,拽

【Key Sentences】

1. The actual knitting elements include the needles, **sinkers**, the needle bed or **frame** that holds the needles, and the yarn carries that lay the yarn in knitting position.

2. In modern knitting, the three major types of needles are the **latch needle**, **spring-bearded needle** and **compound needle**.

3. In the **running position**, the held loop **rests on** the top of the open latch, and **clearing** occurs as the held loop **slips off** the latch and onto the **stem** as the needle moves **upwards**.

4. A **downwards** movement enables the needle hook to **engage** a new piece of yarn, this is known as **feeding**.

5. As the needle continues downwards, the latch is forced to close under the influence of the held loop.

6. **Knockover** occurs as the held loop **disengages** from the needle.

7. Following knockover, loop **pulling** occurs and a new knit loop is formed. The needle must now return to the running position to complete the cycle.

Fig 5.6 shows the diagram for the latch needle in forming a knit loop.

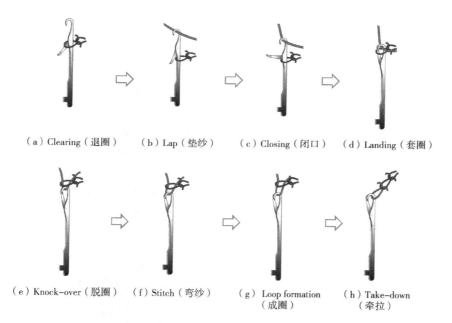

(a) Clearing (退圈)　(b) Lap (垫纱)　(c) Closing (闭口)　(d) Landing (套圈)

(e) Knock-over (脱圈)　(f) Stitch (弯纱)　(g) Loop formation (成圈)　(h) Take-down (牵拉)

Fig 5.6　Diagram for the latch needle in forming a knit loop (舌针成圈过程示意图)

【Text】

纬编组织分为三大类：基础组织、变化组织和花色组织[1]。基本纬编组织包括纬平针、罗纹和双反面组织[2]。纬平针是最简单的一种针织物，也是针织物的基础花型。它在纵向、横向都有很好的拉伸性，而且易于缝制[3]。罗纹也是一种基础组织，通过(正面)线圈纵列和(反面)线圈纵列交替编织而成[4]。罗纹组织在横向有出色的弹性和较好的拉伸回复性，它与同等重量的平针组织相比具有更好的蓬松度[5]。

双反面组织是由线圈横列交替加工而成。因此，双反面针织物的特点是在一个线圈纵列既有正面线圈横列，也有反面线圈横列[6]。双反面针织物在织物纵向有很好的弹性，在织物纵向、横向均有很好的拉伸回复性[7]。变化组织包括变化平纹组织、双罗纹组织,花色组织包括提花组织、集圈组织、衬垫组织和纱罗组织[8]。

【Key Words】

basic construction 基础组织
alternative construction 变化组织
fancy construction 变化组织
jersey [ˈdʒəːzi] construction 纬平纹组织
rib [rib] construction 罗纹组织
purl [pəːl] construction 双反面组织
lengthwise [ˈleŋθwaiz] 纵向地
widthwise 与宽同方向地,纬向地
sew [səu] 缝制

wale 线圈纵列
course 线圈横列
course-wise direction 横向
wale-wise direction 纵向
bulk [纺]蓬松
tuck construction 集圈组织
jacquard construction 提花组织
fleecy construction 衬垫组织
open-work knitting construction 纱罗组织

【Key Sentences】

1. Weft-knitting stitches have three varieties: **basic construction**, **alternative construction** and **fancy construction**.

2. The basic weft-knitting stitches include **jersey**, **rib** and **purl construction**.

3. Jersey knitting fabric has good stretchability in both **lengthwise** and **widthwise**, otherwise, it is easy to **sew**.

Fig 5.7 shows the jesery construction. Fig 5.8 shows the jesery knitting fabric.

4. Rib knitting is the basic construction and made by alternating **wale** of knitted loops.

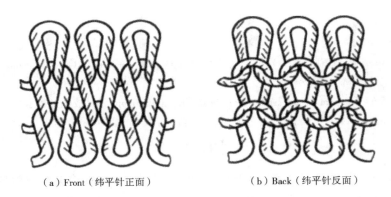

(a) Front(纬平针正面)　　　(b) Back(纬平针反面)

Fig 5.7　Jesery construction(纬平针组织)

5. Rib construction have excellent elasticity and fair stretch recovery in the **course-wise direction** and they provide greater **bulk** than jersey of the same weight.

Fig 5.9 shows 1+1 rib construction, Fig 5.10 shows 1+1 rib knitting fabric, Fig 5.11 shows 2+2 rib construction fabric.

Project Five Knitted Fabrics and Knitting Technology(针织物与针织技术)

Fig 5.8 Jesery knitting fabric(纬平针织物)

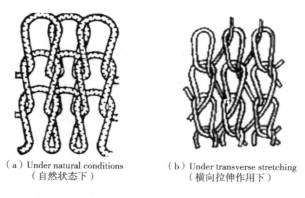

(a) Under natural conditions (b) Under transverse stretching
　　(自然状态下)　　　　　　　(横向拉伸作用下)

Fig 5.9 1+1 rib construction(1+1 罗纹组织)

Fig 5.10 1+1 rib knitting fabric(1+1 罗纹织物)

6. Purl fabrics are characterized with both face knit loops and back knit loops in one wale. Fig 5.12 shows the purl construction, Fig 5.13 shows double rib construction.

163

Fig 5.11 2+2 rib construction fabric(2+2 罗纹织物)

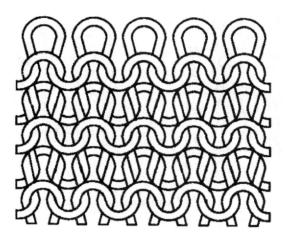

Fig 5.12 Purl construction (双反面组织)

Fig 5.13 Double rib construction(双罗纹组织)

7. Purl fabrics have good elasticity in the **wale-wise direction** and have fair stretch recovery in the **wale-wise** and the course-wise direction. Fig 5.14 and Fig 5.15 show different jacquard construc-

tions and their knitting fabrics.

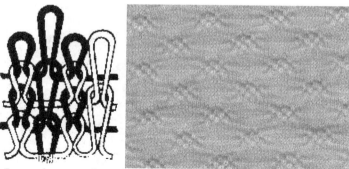

Fig 5.14 Jacquard construction and knitting fabric with uneven sing le-sided structure
[（单面结构不均匀）提花组织与织物]

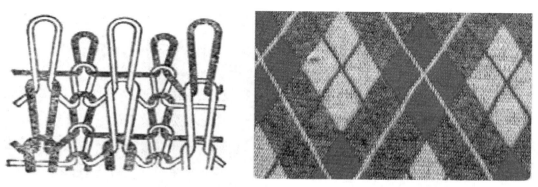

Fig 5.15 Doulde sided jacquard construction and knitting fabric[（双面）提花组织与织物]

8. The alternative constructioninclude **alternative jersery construction** and **double rib construction**. The fancy constructions include **jacquard construction**, **tuck construction**, **fleecy construction** and **open-work knitting construction**.

Fig 5.16 ~ Fig 5.18 shows different tuck constructions and their knitting fabrics. Fig 5.19 and

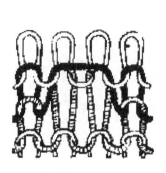   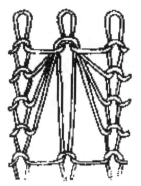

（a）Double needle single row tuck　　（b）Single needle three-row tuck
（双针单列集圈）　　　　　　　　（单针三列集圈）

Fig 5.16 Tuck construction(集圈组织)

Fig 5.20 shows different fleecy constructions and the knitting fabric. Fig 5.21 and Fig 5.22 shows pile/plush construction and its knitting fabrics.

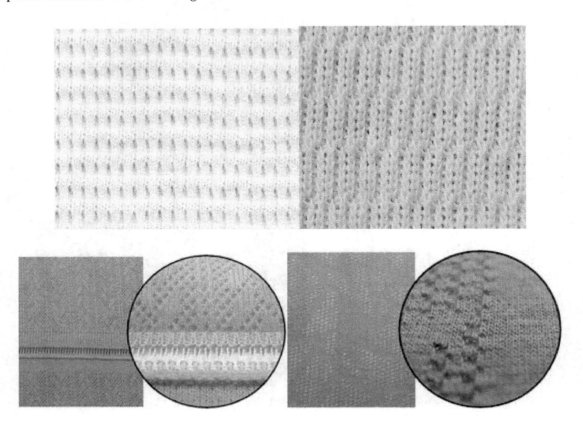

Fig 5.17 One-side tuck construction knitting fabric[(单面)集圈组织织物]

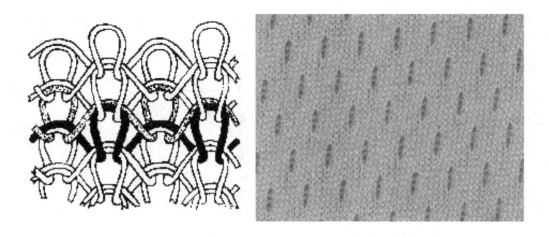

Fig 5.18 Two-sided tuck construction knitting fabric(双面集圈组织)

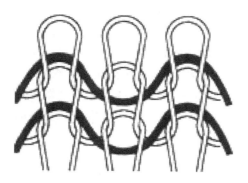

Fig 5.19　Fleecy construction[（平针）衬垫组织]

Fig 5.20　Fleecy construction and its knitting fabric[（三线）衬垫组织及其针织物]

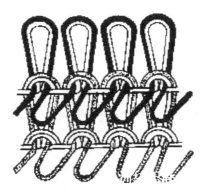

Fig 5.21　Pile/plush construction(毛圈组织)

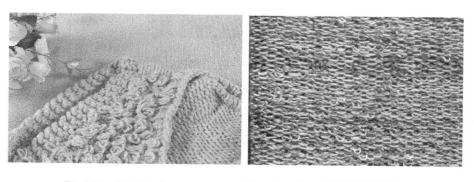

Fig 5.22　Pile/plush construction and knitting fabric(毛圈组织织物)

【Text】

纬编是将纱线形成的线圈沿着织物的宽度方向或以一个圆形形成织物的加工过程[1]。大多数纬编织物是用圆形针织机(简称圆纬机)生产的,圆形针织机上一系列的织针沿着一个圆环排列[2]。(圆纬机)加工的针织物为圆形管状物,如果需要平幅织物,可以将管状针织物剪开[3]。纬编织物也可以在横机(织针直线排列)上编织。使用多种类型的控制如计算机控制,圆型针织机和横机均可生产出复杂花型织物[4]。

【Key Words】

construction process 加工过程
circle ['sə:kl] 圆形,圆周,圆圈
a considerable amount of 相当数量的,相当多的
circular knitting machine 圆形针织机 (圆纬机)
circumference [sə'kʌmfərəns] 圆周
tube 管状物
fiat fabric 平幅织物
flat bed machines [针]横机

【Key Sentences】

1. Weft or filling knitting is a **construction process** in which the fabric is made by yarn forming loops across the width of the fabric or around a **circle**.

2. **A considerable amount of** filling knit fabric is made on a **circular knitting machine**, in which a series of needles is arranged around the **circumference** of a circle.

Fig 5.23 shows the circular knitting machine.

Fig 5.23 Circular knitting machine(圆纬机)

3. Fabric may be made in the shape of a **tube**. If a fiat fabric is desired, the tube can be cut open.

4. Both circular and **flat bed machines** can produce highly patterned fabrics through the use of various types of controls such as computer controls.

Fig 5.24 shows the flat bed machines. Fig 5.25 shows the fully automatic computer flat bed machines.

Project Five　Knitted Fabrics and Knitting Technology(针织物与针织技术)◀

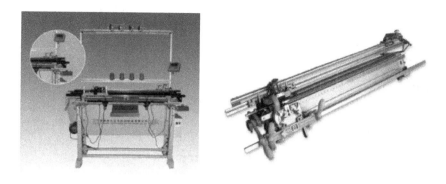

Fig 5.24　Flat bed machines(手摇横机)

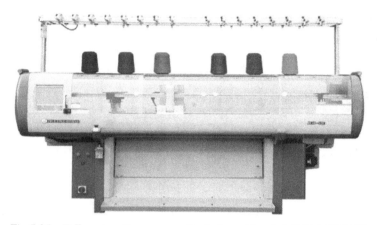

Fig 5.25　Fully automatic computer flat bed machines(全自动电脑横机)

【Text】

经编织物可采用钩针和槽针进行加工。经编机的编织需要各成圈机件的配合,经编机作用的织针类型决定其成圈运动的配合形式[1]。为适应高速和高质量编织的需要,现代经编机大多使用槽针[2]。不管是钩针还是槽针,其成圈步骤基本与舌针类似,也经过退圈、垫纱、闭口、套圈、弯纱、脱圈、成圈和牵拉等过程。

经编织物紧密,没有拉伸弹性,不像纬编织物那么蓬松。两种重要的经编织物是经平纹和拉舍尔针织物[3]。经平纹组织是(经编织物中)最简单的花型,也是应用最广的[4]。经编与纬编的不同之处在于它所形成的线圈是沿垂直或沿经纱方向排列的,由相邻的纱线相互串套而成[5]。经编所用的机器看上去有些像织机[6]。从根本上说,经编是生产平幅织物的加工系统。织物具有直的侧边,生产速度高,产量较大[7]。尽管与纬编织物相比,伸长性和拉伸性较低,但经编织物仍具有一定的伸长性和拉伸性[8]。

【Key Words】

cooperation [kəuˌɔpəˈreiʃn] 合作,协作,协助

coordination [kəuˌɔːdiˈneiʃn] form 配合形式

slot needle 槽针
stretch 延伸性，弹性
bulky ［纺］蓬松的
major importance 至关重要
tricot 经平纹（针织物）
raschel 拉舍尔（针织物）

be adapted from 改编自，来源于
side by side 并肩（排列），此处可译为相邻（排列）
interlop 串套
essentially [i'senʃəli] 本质上；根本上
stretchability 拉伸性；弹性

【Key Sentences】

1. The weaving of the warp knitting machine requires the **cooperation** of the various looping parts, and the type of the knitting needle on the warp knitting machine determines the **coordination form** of the stitching motion.

Fig 5.26 shows the different warp knitting process of knitting needles.

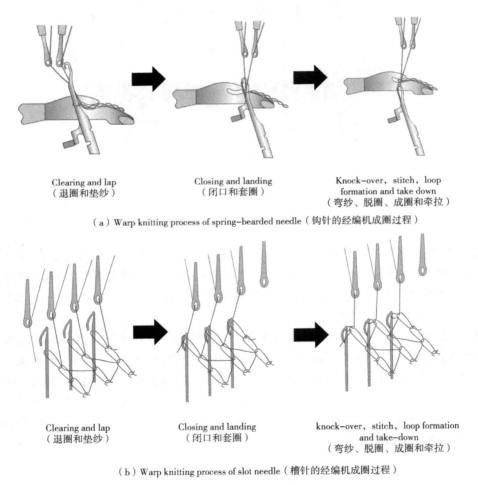

Clearing and lap（退圈和垫纱）　　Closing and landing（闭口和套圈）　　Knock-over, stitch, loop formation and take down（弯纱、脱圈、成圈和牵拉）

（a）Warp knitting process of spring-bearded needle（钩针的经编机成圈过程）

Clearing and lap（退圈和垫纱）　　Closing and landing（闭口和套圈）　　knock-over, stitch, loop formation and take-down（弯纱、脱圈、成圈和牵拉）

（b）Warp knitting process of slot needle（槽针的经编机成圈过程）

Fig 5.26　Different warp knitting process of knitting needles（不同经编机成圈过程）

2. In order to meet the needs of high speed and high quality weaving, modern warp knitting ma-

chines mostly use **slot needle**s.

3. The two types of warp-knit cloth of **major importance** are **tricot** and **raschel**.

4. Tricot fabrics are the simplest pattern and the most widely used.

Fig 5.27 shows the tricot.

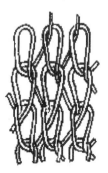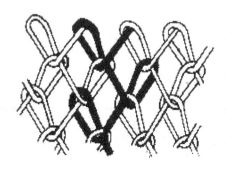

Fig 5.27 Tricot(经平组织)

5. Warp knitting differs from weft or filling knitting in that the loops are formed in a vertical or warp-wise direction and yarns lying **side by side** are **interloped**.

6. Machines used for warp knitting tend to look somewhat like weaving machines.

Fig 5.28 shows the warp knitting machine.

7. **Essentially**, warp knitting fabric is flat and has straight side edges. Also it is manufactured rapidly and in large quantities.

8. Warp knit fabrics have some elongation or **stretchability**, although it may be less than filling knit fabric.

Fig 5.28 Warp knitting machine(经编机)

Part 2 *Trying*

1. According to the performance characteristics of woven fabrics and knitted fabrics, selecte suitable fabrics for the following textile and garment products: underwear, bed linen, pillowcases, suits,

T-shirts, jeans, sweatshirts, socks, sweaters.

	Woven fabric	Knitted fabric
Textile and garment products		
Reasons for the selection		

2. According to the knowledge you learned in the textbook, please describe briefly the difference between warp knitting and weft knitting processing technology?

_____
_____
_____

3. Under the guidance of the training teacher, learn the operation of flat bed machine in the training factory.

Part 3 *Thinking*

1. Combined with the weaving technology studied earlier, analyze the difference of the process between woven fabrics and knitted fabrics.

_____
_____
_____

2. Think about why the warp knitted fabric is less used than weft knitting? Go to the market and buy a piece of warp knitted fabric and a piece of weft knitting fabric, please do a simple comparison for them.

_____
_____
_____

Project Six　Nonwoven Fabrics and Processing Technology(非织造物与非织造技术)

Task One　Nonwoven Fabrics(非织造物)

【Text】

　　非织造织物通常指不通过常规方法如机织、针织等生产的织物[1]。在非织造织物中,纤维网通过各种方式黏合形成整体织物结构[2]。尽管国际标准组织于1988年公布了一项(非织造布的)定义(ISO 9092:1988),但仍没有国际公认的非织造布定义[3]。许多权威机构认为ISO 9092的定义非常狭窄,因为在定义中,毛毡、针刺织物和缝编材料可能没有被严格描述[4]。

　　"非织造织物是通过机械、化学、热或溶剂装置及其组合实现纤维之间的黏合、互锁或黏合互锁相结合产生的织物结构[5]。非织造织物不包括织造、针织或簇绒的纸或织物。"这是美国材料试验协会(ASTM D 1117—80)对非织造织物的定义[6]。这个定义包括许多重要的面料,大多数人认为这些面料是非织造布,但是被ISO 9092排除在外[7]。

【Key Words】

nonwoven ['nɔn'wəuvən] fabric　非织造织物
conventional [kən'venʃnl]　传统的,习用的
integral ['intigrəl]　完整的
internationally [ˌintə'næʃnəli]　国际性地,国际上地
definition [ˌdefi'niʃn]　定义
International Standards Organization　国际标准组织
authority [ɔː'θɔrəti]　权威机构或部门
stitch-bonded　缝编的
interlock [ˌintə'lɔk]　连锁,互锁
accomplish [ə'kʌmpliʃ]　完成,实现
thereof [ˌðeər'ɔv]　其,它的,关于那,由此　tuft　用丛毛装饰,使分成簇(或束)
American Society for Testing Material　美国材料试验协会
exclude [ik'sklu:d]　不包括,不放在考虑之列

【Key Sentences】

1. **Nonwoven fabrics** usually refer to those fabrics that have not been produced by **conventional** methods such as weaving, knitting, and so on.

2. Fig 6.1 shows the transformation of collective morphology about woven fabric and nonwoven fabric.

3. However, there is no **internationally** agreed **definition** of nonwovens, in spite of the fact that the **International Standards Organization** published a definition in 1988 (ISO 9092:1988).

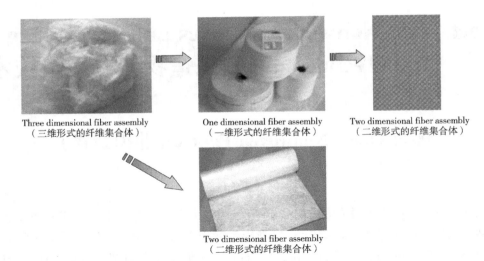

Fig 6.1   Transformation of collective morphology about woven fabric and nonwoven fabric
（机织物与非织造物中纤维集合形态的转变）

4. Many **authorities** believe that the definition of nonwovens in ISO 9092 is very narrow, because felts, needled fabrics, and **stitch-bonded** materials may not strictly be described in the definition.

5. Nonwoven fabric is a textile structure produced by bonding, **interlocking** and their combination of fibers, which are, **accomplished** by mechanical, chemical, thermal or solvent means and combinations **thereof**.

Fig 6.2 shows the microstructure of woven fabric and knitted fabric. Fig 6.3 shows the microstructure of spunlanced and bonded nonwoven fabrics.

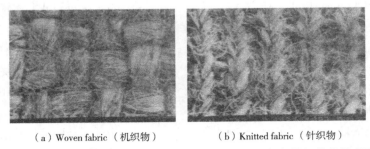

（a）Woven fabric（机织物）　　　（b）Knitted fabric（针织物）

Fig 6.2   Microstructure of woven fabric and knitted fabric（机织物与针织物的微观形态结构）

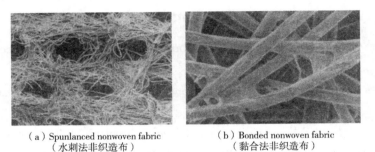

（a）Spunlanced nonwoven fabric　　　（b）Bonded nonwoven fabric
（水刺法非织造布）　　　　　　　　　（黏合法非织造布）

Fig 6.3   Microstructure of spunlanced and bonded nonwoven fabrics
（水刺法非织造布与粘合法非织造布的微观形态结构）

6. This is the definition of nonwoven fabrics by the **American Society for Testing Materials** (ASTM D 1117-80).

7. This definition includes many important fabrics which most people regard as nonwovens, but **excluded** by ISO 9092.

【Text】

非织造织物的主要优点之一是其制造速度,其通常比所有其他形式的织物的生产快得多,例如,纺黏可以比织造快 2000 倍[1]。这就意味着非织造织物的加工非常经济,并且成本非常低[2]。尽管如此,非织造织物也是功能多样化的材料。具有多种不同物理形态、多种不同性能的(非织造布)可以通过不同纤维结合不同的工艺技术和黏合剂来生产[3]。

几乎所有的非织造织物都是通过两个主要步骤制造的。非织造布加工的第一步是纤维准备工作,加工纤维以形成纤维网[4]。纤维成网的方式有许多种,每种方式都会在最终织物中产生自己的特殊特征[5]。非织造布加工的第二步是纤维黏合,将纤维网内纤维黏合在一起。同样,有许多不同的黏合方式,它们会对成品织物产生更大的影响[6]。几乎每种成网方式都可以与每种黏合方式结合,因此,非织造织物的可能生产线,以及最终的非织造织物产品是巨大的[7]。非织造织物通常根据两个主要制造步骤命名。干法成网非织造布、湿法成网非织造布、熔喷法非织造布、针刺法非织造布、纺粘法非织造布、缝编法非织造布和水刺法非织造布都是典型的非织造织物[8]。

【Key Words】

spunbonding 纺粘
economical [ˌiːkəˈnɔmikl] 经济的,节约的,合算的
versatile [ˈvəːsətail] 多才多艺的,多功能的
binder [ˈbaində(r)] 黏合剂,结合剂
physical form 物理形式
fiber web 纤维网
web formation 成网
characteristic [ˌkærəktəˈristik] 特征,特点
equally 同样地

manufacturing line 生产线
bonding of the web 纤维网黏合
enormous [iˈnɔːməs] 巨大的,庞大的
dry-laid 干法成网
wet-laid 湿法成网
melt-blown 熔喷法
needlepunched 针刺法
spunbonded 纺粘法
stitch-bonded 缝编法
spunlanced 水刺法
typical [ˈtipikl] 典型的,代表性的

【Key Sentences】

1. One of the major advavtages of nonwoven fabrics is their speed of manufacture, which is usually much faster than all other forms of fabrication, for example, the production speed of **spunbonding** can be 2,000 times faster than weaving.

2. This means that the manufacture of nonwoven fabrics is very **economical**, and the cost is very low.

3. A variety of different physical forms with a wide range of properties can be produced from different fibers with different technique and **binders**.

4. The first step is the preparation of the fibers, in which fiber is processed to form a **fiber web**.

5. There are a number of different ways of **web formation**, and each way will produce its own particular **characteristics** in the final fabric.

6. **Equally**, there are a number of different ways of **bonding of the web**, and they will have an even bigger effect on the finished fabric.

7. Almost every web formation way can be combined with every bonding way, so that the possible **manufacturing lines** of nonwoven fabric, and therefore the final nonwoven fabric products are **enormous**.

8. **Dry-laid**, **wet-laid**, **melt-blown**, **needlepunched**, **spunbonded**, **stitch-bonded**, and **spunlaced** are typical nonwoven fabrics.

【Text】

非织造布可以应用在很多领域[1]。主要的应用领域包括：服装、制鞋用非织造产品：黏合衬、衬里、垫肩、喷胶棉、内衣裤、浴衣；鞋垫、合成革（鞋面）[2]；家用非织造布：被胎、枕芯、床垫、地毯、墙布、沙发布、包装用品、餐巾纸、清洁布[3]；医用及卫生材料：手术衣帽、口罩、包扎材料、纱布、医用床单、床罩、婴儿尿布、卫生巾[4]；土木建筑用非织造布：水利、铁路、公路、机场、运动场用土工布，土工隔栅、渗水管、防水材料[5]；汽车用非织造布：汽车车厢垫材、汽车保温夹层、汽车内地毯、汽车外衣、顶篷、座椅外套等装饰材料[6]；工业用非织造布：过滤材料（液体过滤、防毒、耐高温过滤），篷盖材料（防护盖布），绝缘材料（隔热防寒、电绝缘），各种吸附材料（吸水材料、吸油材料），工业运输带及增强材料等[7]；农业、园艺用非织造布：保温、覆盖、遮光、防病虫害用非织造布[8]；军事、国防用非织造布：航天、船舶用材，高温隔热材料，军用帐篷，军用地图，降落伞用材，飞行员御寒服[9]；另外，也可以用于加工包装袋、油画布、书法纸、钞票纸、玩具材料标签、商标、人造花等[10]。

【Key Words】

field 领域
adhesive lining 黏合衬
lining 衬里
pad shoulder 垫肩
spray-bonded cotton 喷胶棉
bath clothes 浴衣
insole [ˈinsəul] 鞋垫
instep [ˈinstep] 鞋面
quilt [kwilt] 被胎
pillow [ˈpiləu] 枕芯
mattress [ˈmætrəs] 床垫

napkin [ˈnæpkin] 餐巾纸
sanitary [ˈsænətri] 卫生的，公共卫生的
surgical [ˈsəːdʒikl] 外科的，外科手术的
mask 口罩
bandage [ˈbændidʒ] 绷带
gauze [gɔːz] 纱布（包扎伤口用）
sheet [ʃiːt] 床单，被单
bedspread 床罩
diaper [ˈdaipə(r)] （婴儿的）尿布，尿片
sanitary napkin 卫生巾
civil construction 土木建筑

geotextile 土工布
water conservancy [kən'sə:vənsi] 水利
stadium ['steidiəm] 体育场，运动场
geogrid 土工格栅
water seepage ['si:pidʒ]（渗透）
pipe 渗水管
waterproof 防水的
compartment [kəm'pɑ:tmənt] 车厢
cushion ['kuʃn] 垫材
decorative ['dekərətiv] 装饰性的，作装饰用的
mezzanine ['mezəni:n] 夹层
toxic ['tɔksik] 有毒的
high temperature resistant filtration 耐高温过滤材料
horticulture ['hɔ:tikʌltʃə(r)] 园艺学，园艺
black-out 遮光
pest control 防虫害
military ['milətri] 军事的
national defense 国防
aerospace ['eərəuspeis] 航空航天
vessel ['vesl] 船舶
tent [tent] 帐篷
parachute ['pærəʃu:t] 降落伞
pilot ['pailət] 飞行员
canvas 画布
calligraphy [kə'ligrəfi] 书法，书法艺术
banknote 钞票
trademark 商标
artificial [,ɑ:ti'fiʃl] 人造的

【Key Sentences】

1. Nonwovens can be used in many **fields**.

2. Nonwoven fabrics used in clothing and shoes: adhesive lining, lining, pad shoulder, **spray-bonded cotton**, underwear, **bath clothes**, **insoles**, synthetic leather (**instep**).

3. Household nonwoven fabrics: **quilt**, **pillow**, **mattress**, carpet, wall cloth, sofa cloth, packaging supplies, **napkin**, cleaning cloth.

4. Medical and **sanitary** materials: **surgical** clothing and cap, breathing mask, **bandage** material, **gauze**, medical **sheets**, **bedspreads**, baby **diapers**, **sanitary napkins**.

Fig 6.4 shows some nonwoven fabics used in clothing and shoes and household nonwoven fabrics.

5. Nonwoven fabrics for **civil construction**: **geotextiles** used in **water conservancy**, railways, highways, airports, and **stadiums**, **geogrid**, **water seepage pipes**, **waterproof** materials.

6. Nonwoven fabrics for automobiles: car **compartment cushions**, car insulation **mezzanines**, carpets in cars, car coats, roofs, seat coats and other **decorative** materials.

Fig 6.5 shows some nonwoven fabrics for medical and sanitary materials, civil construction and automobiles.

7. Nonwoven fabrics for industrial use: filtration materials (liquid filtration, anti-**toxic** filtration, **high temperature resistant filtration**), cover materials (protective cover cloth), a variety of insulation materials (heat insulation, cold insulation and electric insulation), a variety of adsorption materials (water absorbing materials and oil absorbing material), industrial transport belt and enhanced materials.

Fig 6.6 shows some nonwoven fabrics for industry and agriculture.

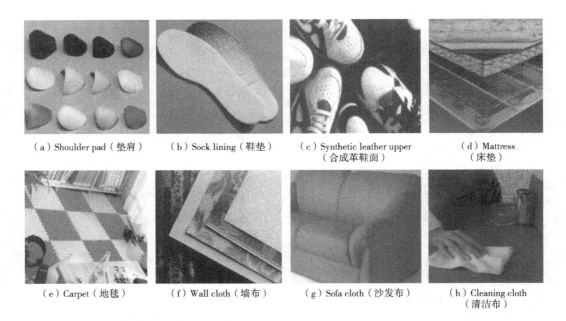

(a) Shoulder pad（垫肩） (b) Sock lining（鞋垫） (c) Synthetic leather upper（合成革鞋面） (d) Mattress（床垫）

(e) Carpet（地毯） (f) Wall cloth（墙布） (g) Sofa cloth（沙发布） (h) Cleaning cloth（清洁布）

Fig 6.4 Nonwoven fabrics used in clothing and shoes and household nonwoven fabrics
（用于服装、鞋及家居用非织造物）

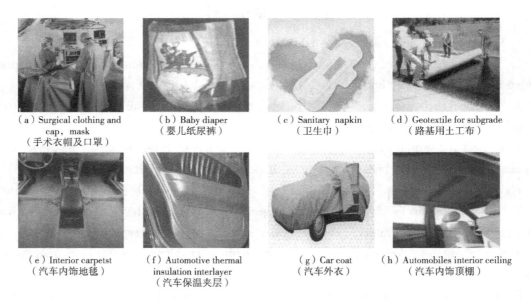

(a) Surgical clothing and cap, mask（手术衣帽及口罩） (b) Baby diaper（婴儿纸尿裤） (c) Sanitary napkin（卫生巾） (d) Geotextile for subgrade（路基用土工布）

(e) Interior carpetst（汽车内饰地毯） (f) Automotive thermal insulation interlayer（汽车保温夹层） (g) Car coat（汽车外衣） (h) Automobiles interior ceiling（汽车内饰顶棚）

Fig 6.5 Nonwoven fabrics for medical and sanitary materials, civil construction and automobices
（用于医疗、卫生、城建和汽车领域的非织造物）

8. Nonwoven fabrics for agriculture and **horticulture**: nonwovens for thermal insulation, cover, **black-out**, pest control.

9. Nonwoven fabrics for **military**, **national defense**: **aerospace** and **vessel** materials, high temperature insulation materials, military **tents**, military maps, **parachute** materials, cold protective clothing of **pilot**.

Project Six Nonwoven Fabrics and Processing Technology(非织造物与非织造技术)

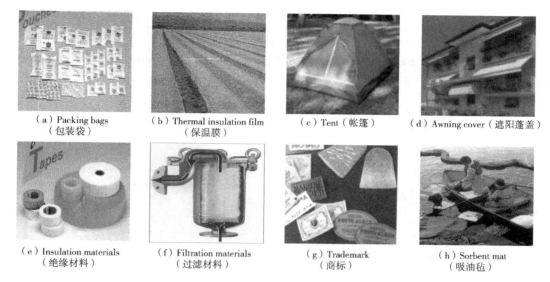

(a) Packing bags (包装袋)　　(b) Thermal insulation film (保温膜)　　(c) Tent (帐篷)　　(d) Awning cover (遮阳蓬盖)

(e) Insulation materials (绝缘材料)　　(f) Filtration materials (过滤材料)　　(g) Trademark (商标)　　(h) Sorbent mat (吸油毡)

Fig 6.6　Nonwoven fabrics for industry and agriculture(工业用和农用非织造物)

10. In addition, it can also be used to make bags, oil **canvas**, **calligraphy** paper, **banknote** paper, toy material labels, **trademarks**, **artificial** flowers and so on.

Part 2　*Trying*

1. Select three kinds of nonwoven fabrics, woven fabrics and knitted fabrics with the same raw material, and compare them simply in structure and performance.

	Nonwoven fabrics	Knitted fabrics	Woven fabrics
Structure			
Performance			

2. Are there any similarities of the process between nonwoven fabric and the woven fabric? If so, what are the similarities?

179

Part 3 *Thinking*

1. Go to the market and buy some nonwoven fabrics, do you find nonwoven fabics more used in the clothing? Why?

_____
_____
_____

2. In terms of economic costs, what are the advantages of nonwoven technology compared with knitting technology and woven technology?

_____
_____
_____

Task Two   Netting Technology(纤维成网技术)

【Text】

干法成网是纤维成网最常用的一种方法。干法成网的方式有三种：平行铺网、交叉铺网及气流成网[1]。它们占非织造布总产量的一半不到[2]。三种干法成型工艺与传统的纺纱加工相似。

平行铺网与交叉铺网首先都需要将纤维制成纤维网(加工过程与纺纱过程基本相同)。在平行铺网中,每层纤维通过梳理生产。梳理的原理和梳理机的类型与纱线制造非常相似[3]。每层纤维网的质量和强度通常太低[4]。几层纤维网平行放置,因此,命名为"平行铺网"[5]。另外,通过将多个梳理纤维层彼此重叠形成纤维网,可以提高均匀性[6]。来自每个梳棉机的纤维层落在传送带上,形成质量是每层纤维网三倍的纤维网[7]。

在非织造布中,沿着纤维网的方向称为机器方向(纵向),与纵向垂直的方向称为横向[8]。然而,平行铺网有两个缺点。第一个缺点是横向(宽度方向)强度低,这个缺点可以通过特殊设计的梳棉机来改善,即在梳棉机上加一个随机道夫和不规则罗拉[9]。另外一个缺点是最终非织造织物的宽度受限制。非织造织物的宽度不能宽于梳理纤维网,而现在纺织品市场需要越来越宽的织物。

【Key Words】

dry-laying 干法成网
parallel-laid ['pærəlel] 平行铺网
cross-laid 交叉铺网
air-laid 气流成网
account for (在数量、比例上)占……
mass [物理学] 质量

normally ['nɔːməli] 正常地,通常地,一般地
additionally [əˈdiʃənəli] 此外
uniformity [ˌjuːniˈfɔːməti] 均匀性
conveyor [kənˈveiə(r)] 传送带
drawback 缺点,不理条件

weakness ['wi:knəs] 弱点,缺点
cross direction=width direction 横向
alter ['ɔ:ltə(r)] 改变,更改
randomising 随机的
scrambling 快速的

【Key Sentences】

1. There are three methods in **dry-laying**: **parallel-laid**, **cross-laid** and **air-laid**.
2. They **account for** slightly less than half the total nonwoven production.
3. The principles of carding and the types of carding machines are quite similar to yarn manufacture.
4. The **mass** and strength of each layer of fiber webs are **normally** too low.
5. Several layers of fibers are laid parallel, so named "parallel-laid".

Fig 6.7 shows parallel-laid(series connection style). Fig 6.8 shows parallel-laid(paprallel connection style).

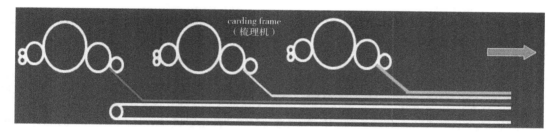

Fig 6.7 Parallel-laid(series connection style)[平行铺网(串联式)]

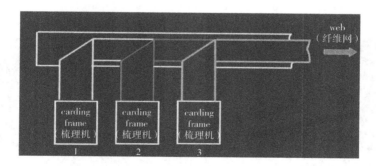

Fig 6.8 Parallel-laid(paprallel connection style)[平行铺网(并联式)]

6. **Additionally**, the **uniformity** can be increased by laying several carded fiber layers over each other to form the fiber webs.
7. The layers of fibers from each card fall onto the **conveyor** forming a web with three times the mass of each layer of fibers.
8. In nonwovens, the direction along the web is called the "machine direction" and the direction perpendicular to the machine direction is called the "cross direction".

9. The first drawback is the **weakness** in **width direction**, that can be **improved** by the special designed cards which are added a **randomising** doffer and **scrambling** rollers.

10. The width of the ultimate nonwoven fabrics cannot be wider than the carded layer of fiber webs, while the textile market demands wider and wider fabrics.

【Text】

在交叉铺网中,每层纤维网也通过梳理。然而,梳棉机与主输送机成直角放置,梳理的纤维层沿主传送带来回穿过[1]。交叉铺网中不存在横向强度低和宽度受限的问题[2]。交叉铺网存在另外两个主要缺点:一个是倾向于在边缘比在中间放置更重的网[3],这个缺点可以采用在中心运行很慢同时在边缘运行得更快的机构来改善[4];另一个缺点是试图将交叉铺网机的(纤维网)输入速度与梳理纤维网的速度相匹配[5]。由于各种原因,交叉铺网的(纤维网)输入速度受到限制,并且必须降低梳理纤维网的速度以匹配它[6]。

【Key Words】

traverse [trəˈvəːs] 横越,穿过　　backwards and forwards 来回
backwards 向后　　mechanism [ˈmekənizəm] 机制
forwards 向前　　input speed (纤网的)输入速度

【Key Sentences】

1. In corss-laid, the cards are placed perpendicular to the main conveyor, and the carded layers of fibers are **traversed backwards** a**nd forwards** across the main conveyor.

Fig 6.9 shows the cross-laid (vertical clamping style), Fig 6.10 shows the cross-laid (four conveyor lattices), Fig 6.11 shows the cross-laid (two lattices clamping).

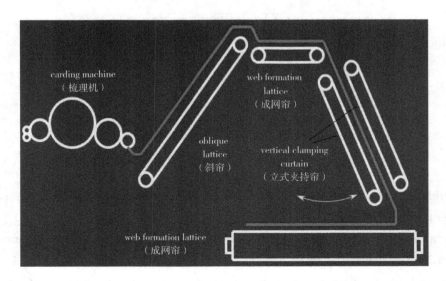

Fig 6.9　Cross-laid (vertical clamping style)[交叉铺网(立式夹持铺叠成网)]

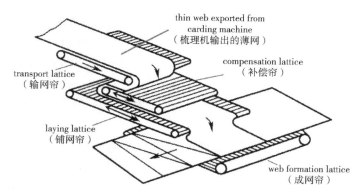

Fig 6.10　Cross-laid (four conveyor lattices)[交叉铺网(四帘式)]

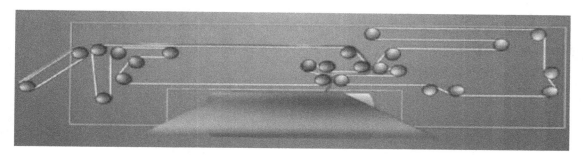

Fig 6.11　Cross-laid (two lattices clamping)[交叉铺网(双帘夹持)]

2. The problem of the weakness of strength in cross direction and the limited width do not exist in cross-laid.

3. One drawback is that cross-laid tend to lay the web heavier at the edges than in the middle.

4. This drawback can be improved by running the traversing **mechanism** rather slower in the center and more rapidly at the edges.

5. The other drawback is trying to match the **input speed** of the cross layer with the card web speed.

6. For various reasons, the input speed of the cross layer is limited and the speed of the card web has to be reduced to match it.

【Text】

气流成网方法一步即可产生最终的纤维网而不需要首先加工纤维网(与平行铺网、交叉铺网相比)。它还能够以高生产速度运行。但它的缺点类似于平行铺网方法,因为最终纤维网的宽度是有限的[1]。气流成网的基本原理如下:从开棉机/混棉机开松的纤维被送入喂料斗的后部,喂料斗向喂棉罗拉提供均匀的纤维棉层[2]。然后纤维被锯齿辊带走,锯齿辊高速旋转[3]。强空气流将纤维从辊的表面移走并将它们带到传送带上,形成纤维网[4]。剥棉板防止纤维在罗拉周围再循环,负压气流有助于纤维稳定成形区域[5]。常用的气流成网方式有五种,分别是自由飘落式、压入式、抽吸式、封闭循环式以及压吸结合式。

【Key Words】

be capable ['keipəbl] of 能够
in that 因为
hopper ['hɔpə(r)] 送料斗,漏斗
toothed roller 锯齿罗拉
revolve [ri'vɔlv] 使旋转
dislodge [dis'lɔdʒ] 强行去除,取出,移动
stripping rail 剥棉板
recirculate [ri'sə:kjuleit] 再循环
negative pressure airflow 负压气流
stabilize ['steibəlaiz] (使)稳定,(使)稳固
free fall 自由飘落式
press-in 压入式
suction 抽吸式
closed-loop 封闭循环式
press-suction combination 压吸结合式

【Key Sentences】

1. The drawback of the air-laid is similar to the parallel-laid method **in that** the width of the final web is limited.

Fig 6.12 shows the diagram of air-laid.

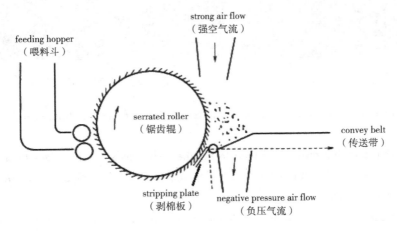

Fig 6.12　Digram of the air-laid(气流成网原理)

2. Opened fiber from the opening or blending section is fed into the back of **hopper**, which delivers a uniform sheet of fibers to the feed rollers.

3. The fiber is then taken by the **toothed roller**, which is **revolving** at high speed.

4. A strong air stream **dislodges** the fibers from the surface of roller and carries them onto the conveyor on which the web is formed.

5. The **stripping rail** prevents fibers from **recirculating** round the cylinder and the **negative pressure airflow** helps the fiber to **stabilize** in the formation zone.

6. There are five common ways of air-laid: **free fall, press-in, suction, closed-loop** and **press-suction combination**.

Fig 6.13 shows different air-laid ways.

Project Six Nonwoven Fabrics and Processing Technology（非织造物与非织造技术）

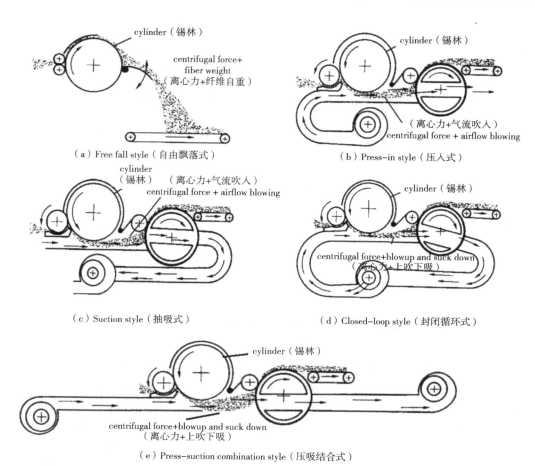

Fig 6.13 Air-laid ways（气流成网方式）

【Text】

  湿法成网的主要优点是生产速度高。湿法成网是造纸发展而来的[1]。根据纺织品标准，纺织纤维的切断非常短，为6~20mm，但与木浆纤维（纸浆的常用原料）相比，这些纤维很长[2]。如果纺织纤维分散在水中，稀释率必须足够大才可防止纤维聚集[3]。所需的稀释率必须大约是纸张所需稀释率的10倍，这意味着只能使用专门形式的造纸机[4]。实际上，最常使用的是纺织纤维与木浆的混合物作为原料，不仅降低了必要的稀释率，而且还使原料涂层大量减少[5]。

  一致认为含有50%纺织纤维和50%木浆的材料为非织造材料，但任何木浆含量进一步增加都会形成纤维增强纸[6]。许多产品恰好使用50%的木浆。湿法成网非织造布约占市场总量的10%，但这一比例趋于下降。它们广泛用于一次性产品中，例如，在医院中用作帷帘、手术服，有时用作床单和一次性尿布[7]。

【Key Words】

papermaking ['peipəmeikiŋ] 造纸，造纸业，造纸学

in comparison with 与…比较起来

wood pulp 木质纸浆

disperse into　分散于…中
dilution [dai'lju:ʃn]　稀释
aggregate ['ægrigət]　使聚集，使积聚
turn out to be　结果是，原来是，证明是
content ['kɔntent]　含量
result in　引起，导致
represent [ˌrepri'zent] about　占……

decline [di'klain]　下降
disposable [di'spəuzəbl]　一次性的，可任意处理的
drapes　帷帘，帷幕
gown [gaun]　长外衣，此处解释为手术服
sheets [ʃi:ts]　被单，床单
nappies　尿布

【Key Sentences】

1. The wet-laying process develop from **papermaking**.

2. Textile fibers are cut very short according to textile standards (6~20mm), but at the same time, these are very long **in comparison with wood pulp**, the usual raw material for paper.

3. If the fibers are **dispersed into** water, the rate of **dilution** has to be great enough to prevent the fibers from **aggregating**.

4. The required dilution rate **turns out to be** roughly ten times of that required for paper, which means that only specialized forms of paper machines can be used.

5. In fact, the mixture of textile fibers and wood pulp is most commonly used as raw materials, not only reducing the necessary dilution rate but also causing to a big reduction in the coat of the raw material.

6. **It has been agreed that** a material containing 50% textile fiber and 50% wood pulp is a non-woven, but any further increase in the wood pulp **content results in** a fiber-reinforced paper.

7. Wet-laid nonwovens are used widely in **disposable** products, for example, in hospitals as **drapes**, **gowns**, sometimes as **sheets**, and in disposable **nappies**.

【Text】

生产湿法非织造布，首先需要将（纤维）原料和化学助剂制成悬浮浆[1]。悬浮浆的组成成分包括纤维、分散剂、黏合剂（或黏合纤维）、湿增强剂。若纤维原料是纤维素浆粕板，常采用非连续式制浆[2]；若是常规纤维，则采用连续式制浆。常用湿法成网方法有斜网式成网、圆网式成网以及复合式成网[3]。

【Key Words】

chemical additive　化学添加剂
suspended [sə'spendid] pulp　悬浮浆
composition　组分，成分
dispersant [dis'pə:sənt]　分散剂
adhesive [əd'hi:s:v]　黏合剂
wet strength agent　湿增强剂

cellulose pulp plate　纤维素浆粕板
storage bucket ['bʌk:t]　储料桶
oblique [ə'bli:k]-laid　斜网式成网
round-laid　圆网式成网
composite-laid　复合式成网

Project Six Nonwoven Fabrics and Processing Technology(非织造物与非织造技术)

【Key Sentences】

1. Preparing the fibe suspended pulp made by raw materials and **chemical additives** is the first step of the prodcution of wet nonwoven fabrics.

2. If the fiber raw material is **cellulose pulp plate**, non-continuous pulping is applied in wet nonwoven process.

Fig 6.14 shows process diagram of non-continuous pulping. Fig 6.15 shows process diagram of continuous pulping.

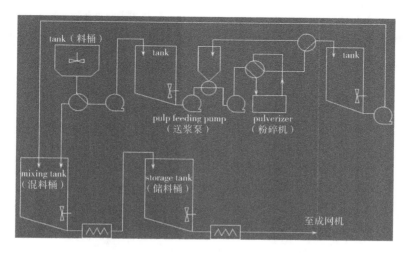

Fig 6.14 Process diagram of non-continuous pulping(非连续式制浆工艺简图)

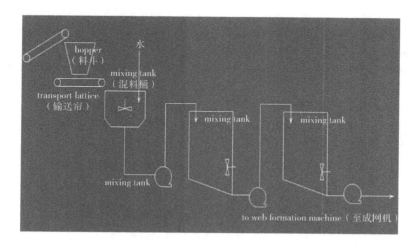

Fig 6.15 Process diagram of continuous pulping(连续式制浆工艺简图)

3. Common wet-laying methods are **oblique-laid**, **round-laid**, as well as **composite-laid**.

Fig 6.16 shows the diagram of oblique-laid process. Fig 6.17 shows the diagram of round-laid process. Fig 6.18 shows the diagram of composite-laid process.

187

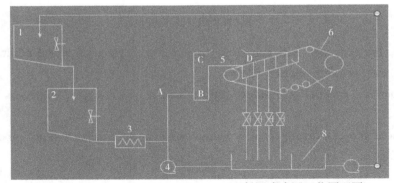

Fig 6.16　Diagram of oblique-laid process(斜网式成网工艺原理图)

1—mixing tank(混料桶)　2—stirringtant(搅料桶)　3—metering pump(计量泵)　4—axial flow pump(轴流泵)　5—web formation tank(成网料桶)　6—web formation lattice(成网帘)　7—water couectiong tank(集水箱)　8—water purifying tank(净水箱)

Fig 6.17　Diagram of round-laid process(圆网式成网工艺原理图)

1—pipe(管道)　2—dispersion rou(分散辊)　3—web formation area(成网区)　4—baffce(挡板)　5—web formation lattice(成网帘)　6—suction box(抽吸箱)　7—water filter tray(滤水盘)　8—rotary drum(回转滚筒)　9—wet web conduction band(湿网导带)　10—overflow adjusting bolt(溢流调节螺栓)

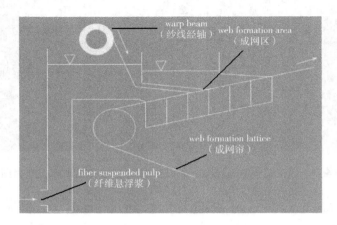

Fig 6.18　Diagram of composite-laid process(复合式成网工艺原理图)

Project Six Nonwoven Fabrics and Processing Technology(非织造物与非织造技术)

Part 2 *Trying*

1. Compare the surface morphology, the advantages and disadvantages of the fiber mets produced by the parallel-laid method, cross-laid method and air-laid method.

	Parallel-laid	Cross-laid	Air-laid
Surface morphology			
Advantages of fiber mets			
Disadvantages of fiber mets			

2. Draw and represent the difference of the fiber arrangement about three kinds of mets produced by the parallel-laid method, cross-laid method and air-laid method.

	Parallel-laid	Cross-laid	Air-laid
Fiber arrangement			

Part 3 *Thinking*

1. Do you know what the principle of papermaking is? Does wet laying have the same principle as papermaking?

2. According to the relevant physical knowledge, what are the requirements of wet laying for fiber properties?

Task Three　Reinforcement Technology(纤网加固技术)

【Text】

将纤维网黏合在一起(加固)有三种常用方法:机械黏合、化学黏合和热黏合[1]。在机械黏合中,使用许多形式的机械处理,例如,刺针、水射流或缝编。因此,对应的机械黏合有三种方法:针刺法、水刺法和缝编法[2]。方法的选择取决于最终产品所需的实际性能、价格和可用设备[3]。

化学黏合通常用黏合剂处理纤维网来实现,黏合剂通过饱和浸渍法、泡沫法、涂层法、印花法、粉末黏合剂散布或通过溶剂基或水基喷洒法施加[4]。一般根据最终用途、耐久性要求、成本或下游加工考虑因素选择黏合剂[5]。

【Key Words】

mechanical bonding　机械黏合法
chemical bonding　化学黏合法
thermal bonding　热黏合法
needle punching　针刺法
hydroentanglement　水刺
stitch bonding　缝编黏合法
end-product　最终产品
available [əˈveiləbl]　可获得的,可用的
carry out　进行,执行
binder [ˈbaində(r)]　黏合剂
apply [əˈplai]　应用,运用
aqueous [ˈeikwiəs] impregnation [ˌimpregˈneiʃn]　饱和浸渍法
foam [fəum] processing　泡沫加工法
coating　涂层法
powder [ˈpaudə(r)] adhesive [ədˈhi:siv]　粉状黏合剂
scatter [ˈskætə(r)]　分开,散开
solvent-based　溶剂型
water-based　水基的,水性的
spray　喷,喷洒
end-use　最终用途
durability [ˌdjuərəˈbiləti]　耐久性,持久性
downstream [ˌdaunˈstri:m]　在下游方向的
consideration [kənˌsidəˈreiʃn]　考虑因素

【Key Sentences】

1. There are three general ways of bonding webs together: **mechanical bonding**, **chemical bonding** and **thermal bonding**.

2. There are three kinds of methods for mechanical bonding: **needle punching**, **hydroentanglement**, and **stitch bonding**.

3. The chosen method depends on the actual properties required in the **end-products**, the price and the **available** equipment.

4. Chemical bonding is usually **carried out** by treating the web with a **binder applied** by **aqueous impregnation**, **foam processing**, **coating**, printing, **powder adhesive scattering** or by a **solvent-based** or **water-based spraying** method.

5. The binders are selected depending on the **end-use**, **durability** requirements, cost or **downstream** processing **considerations**.

Project Six Nonwoven Fabrics and Processing Technology(非织造物与非织造技术)

【Text】

所有常用的化学黏合剂包括丙烯酸、丙烯酸共聚物、丁苯胶乳、乙酸乙烯酯乙烯共聚物等[1]。如果树脂(化学黏合剂)是交联型,则黏合剂施加后需要进行干燥和固化[2]。为了将纤维黏合在一起,化学黏合法可以用黏合剂处理整个纤维网或者相互隔离的部分纤维网[3]。虽然可以使用许多不同的黏合剂;但现今仅使用合成乳胶,其中丙烯酸乳胶至少占一半,而苯乙烯—丁二烯乳胶和醋酸乙烯酯乳胶约占四分之一[4]。

当施加黏合剂时,很有必要润湿纤维,否则黏合效果较差[5]。下一步通过蒸发水分并将聚合物颗粒与添加剂一起留在纤维上或纤维之间来干燥乳液[6]。最后阶段是焙烘,并且在该阶段中,使纤维网达到比干燥更高的温度。焙烘的目的是在聚合物颗粒内部和之间形成交联,从而使黏合膜形成良好的黏结强度[7]。

【Key Words】

acrylic [əˈkrilik] 丙烯酸
acrylic copolymer [kəuˈpɔlimə] 丙烯酸共聚物
SBR 丁苯胶乳
vinyl acetate [ˈvainəl] [ˈæsiˌteit] 乙酸乙烯酯
ethylene [ˈeθəˌliːn] 乙烯
copolymer [kəuˈpɔlimə] 乙烯共聚物
cross-linking 交联

with the intention of 怀着……的目的
bonding agent 黏合剂
evaporate [iˈvæpəreit] 使蒸发,使挥发
aqueous [ˈeikwiəs] 水的
component [kəmˈpəunənt] 成分,组分
phase [feiz] 阶段
crosslink [ˈkrɔsliŋk] 交联,交键
cohesive strength 黏结强度

【Key Sentences】

1. All the common chemical adhesives are used, including **acrylic**, **acrylic copolymer**, SBR, **vinyl acetate ethylene copolymer** etc.

2. If the resin is a **cross-linking** type, application of the binder is followed by drying and **curing**.

3. **With the intention of** sticking the fibers together, chemical bonding involves treating either the complete web or isolated portions of the web with a **bonding agent**.

4. Today, however, only synthetic latex is in-uses, with acrylic latex accounting for at least half, and **stryrene-butadiene latex** and **vinyl acetate latex** accounting for about a quarter.

5. When the bonding agent is applied **it is essential that** it wets the fibers, otherwise poor adhesion will be achieved.

6. The next stage is to dry the latex by **evaporating** the **aqueous component** and leaving the polymer particles together with any additives on and between the fibers.

7. The purpose of curing is to develop **crosslinks** both inside and between the polymer particles and so to develop good **cohesive strength** in the binder film.

【Text】

饱和浸渍黏合是用黏合剂润湿整个纤维网,使得所有的纤维都被黏合剂膜所覆盖[1]。为了饱和浸渍纤维网,将其放到黏合剂表面以下。在大多数情况下,纤维网非常开松和薄弱,需要注意保护避免扭曲[2]。液体的作用大大减小了纤维网的厚度。罗拉的挤压进一步降低了纤维网的厚度[3]。因此,饱和浸渍黏合非织造布通常是密实的并且相对较薄[4]。

饱和浸渍黏合的问题之一是使用了太多的水。这不仅增加了干燥成本,而且增加了黏合剂转移的风险。化学品以泡沫的形式应用可以节约用水,这不仅用于非织造布而且也能用于染整工业[5]。黏合剂溶液和一定体积的空气连续通过驱动涡轮机,驱动涡轮机将两种组分击打成一致的泡沫[6];然后将泡沫输送到刮涂式装置的输送网帘上或轧涂式装置的一对挤压罗拉水平辊隙[7];泡沫破碎后被渗透或挤压进纤网内部。泡沫黏合应该仅被认为是饱和黏合的替代方法。印花黏合也能将相同类型的黏合剂施加到网上,但应用在限制区域和固定的图案[8]。黏合剂不能很好地渗透到干燥纤维网中,因此,纤维网首先用水浸渍饱和,然后用印花滚筒或圆网印花机印花[9]。

【Key Words】

saturation [ˌsætʃəˈreiʃn] （达到）饱和状态
bonding agent 黏合剂
distortion [diˈstɔːʃn] 扭曲,变形
squeeze [skwiːz] 挤,榨,捏
compact [kəmˈpækt] （物质）致密的
thin [θin] 薄的
migration [maiˈgreiʃn] 迁移,移居
chemicals [ˈkemiklz] 化学药品
dyeing [ˈdaiiŋ] 染色,染色工艺
finishing [ˈfiniʃiŋ] 整理
driven turbine [ˈtəːbain] 驱动涡轮机
beat [biːt] 接连地击打

consistent [kənˈsistənt] 连续的,一致的
foam 泡沫
horizontal [ˌhɔriˈzɔntl] 水平的
nip 辊隙
impregnate [ˈimpregneit] 使饱和
alternative [ɔːlˈtəːnətiv] 备选的
set 固定的,确定的
pattern 图案;花样
penetrate [ˈpenitreit]……into 渗透
rotary [ˈrəutəri] 旋转的
printing roller 印花滚筒
rotary screen printer 圆网印花机

【Key Sentences】

1. **Saturation bonding** wets the whole web with **bonding agents**, so that all fibers are covered in a film of binder.

Fig 6.19 shows the schematic of saturation bonding.

2. In most cases, the web is very open and weak and care is needed to avoid **distortion**.

3. The thickness is further reduced by the **squeeze** of rollers.

4. **Hence**, saturation bonded nonwoven fabrics are generally **compact** and relatively **thin**.

5. Application of **chemicals** as a foam was developed not only for nonwovens but also for the **dye-**

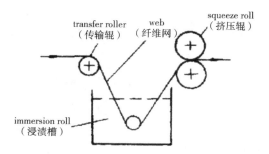

Fig 6.19　Schematic of Saturation bonding(饱和浸渍法)

ing and **finishing** industry as a means of using less water during application.

6. The binder solution and a measured volume of air are passed continuously through a **driven turbine** which **beats** the two components into a **consistent foam**.

Fig 6.20 shows the foam generator.

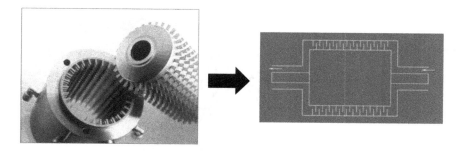

Fig 6.20　Foam generator(泡沫发生器)

7. The foam is then delivered to the conveyor in scratch-coated device or the **horizontal nip** of a pair of squeezing rollers in rolling-coated device.

Fig 6.21 shows scratch-coated method and rolling-coated method. Fig 6.22 shows the crushing foam morphology and fiber bonding morphology.

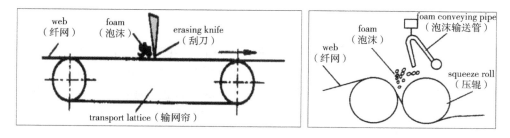

Fig 6.21　Scratch-coated method and rolling-coated method(刮涂式与轧涂式)

8. **Print bonding** involves applying the same types of binder to the web but the application is to limited areas and in a **set pattern**.

Fig 6.23 shows the schematic of print bonding.

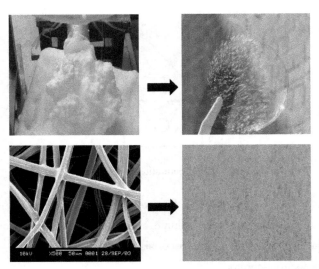

Fig 6.22 Crushing foam morphology and fiber bonding morphology(泡沫破碎形态及纤维黏合形态)

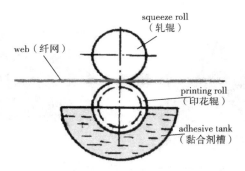

Fig 6.23 Schematic of print bonding(印花黏合原理)

9. The binder does not **penetrate** well **into** the dry web so the web is first saturated with water and then printed with either a **printing roller** or a **rotary screen printer**.

【Text】

  由于游离纤维在未黏合区域中的强烈作用,印花黏合织物的手感更柔软且更有弹性[1]。因纤维在未黏合区域中滑动,它们(强度)也明显弱于饱和浸渍黏合织物。但如果知道纤维长度和纤维取向分布,就可以设计出能够使强度损失最小化的印花黏合图案[2]。印花黏合织物倾向用于注重手感的类似织物[3]。例如,次性防护服、面层材料和擦拭布,特别是家用的。

  类似的乳胶黏合剂也可以应用类似于涂漆中使用的喷枪进行喷洒(施加),喷枪可以通过压缩空气或者无气流(真空)操作[4]。喷涂黏合的最终产品是厚的、开松的和蓬松的织物,广泛用作绗缝织物、羽绒被、一些室内装饰品以及一些过滤介质的填充物[5]。

【Key Words】

owing to 由于
orientation [ˌɔːriənˈteiʃn] 方向
distribution [ˌdistriˈbjuːʃn] 分布
minimize [ˈminimaiz] 使减少到最低限

度；降低
disposable [diˈspəuzəbl] 用后即丢弃的，一次性的
protective 防护的
coverstock 保健卫生用透气织物
wiping 擦，拭，抹，揩
latex [ˈleiteks] 人工合成胶乳（用于制作油漆、黏合剂等）
lofty [纺] 蓬松的，柔软的
filling （枕头、靠垫等的）填充物，填料
quilted [ˈkwiltid] 绗缝的
duvet [ˈduvei] 羽绒被
upholstery [ʌpˈhəulstəri] 室内装饰品
filter media 过滤媒介

【Key Sentences】

1. Print-bonded fabrics are much softer in feel and also much more flexible **owing to** the strong effect of the free fibers in the unbonded areas.

2. If we know the fiber length and the fiber **orientation distribution**, it is possible to design a print pattern which will **minimize** the strength loss.

3. Print-bonded fabrics tend to be used in applications where the textile-like handle is an advantage.

4. Similar **latex** binders may also be applied by spraying, using spray guns similar to those used in painting, which may be either operated by compressed air or be **airless**.

5. The final product of spraying bonding is a thick, open and **lofty** fabric used widely as the **filling** in **quilted** fabrics, for **duvets**, for some **upholstery** and also for some types of **filter media**.

Fig 6.24 shows the quilted print bonding nonwoven fabric.

Fig 6.24　Quilted print bonding nonwoven fabric（绗缝印花黏合非织造物）

【Text】

高分子聚合物材料大都具有热塑性，即加热到一定温度后会软化熔融，变成具有一定流动性的黏流体，冷却后又重新固化，变成固体¹。热黏合充分利用了纤维网（内部纤维）本身的热塑性，或者有时引入外部热熔黏合剂。纤网受热后，部分纤维或热熔粉末软化熔融，纤维间产生黏连，冷却后纤网得到加固而成为热黏合非织造材料²。热黏合一般可以获得面黏合与点黏合两种效果。

通过受控热空气、红外加热器或通过高频或超声波焊接的热轧工艺实现（纤维网）黏合³。出于多种原因，热黏合越来越多地用于取代化学黏合法。比如，热黏合可以高速进行，而化学黏合的速度受干燥和固化阶段的限制；与干燥和固化箱相比，热黏合占用的空间很小；与黏合剂在

干燥过程中蒸发水所需的热量相比,热黏合需要更少的热量,因此,它更节能[4]。

【Key Words】

thermoplastic [θə:məu'plæstik] 热塑(性)塑料
viscous['viskəs] fluid 黏流体
fluidity [flu'idəti] 流动性
hot-melt powder 热熔粉末
area bonding 面黏合
point bonding 点黏合
nature ['neitʃə(r)] 本性,天性
external [ik'stə:nl] 外部的,外面的
hot-melt adhesive 热熔胶
co-spin 共混纺丝
calendar process 热轧工艺
infrared [ˌinfrə'red] heaters 红外加热器
ultrasonic [ˌʌltrə'sɔnik] welding ['weldiŋ] 超声波焊接(黏合)
curing 固化
curing oven 烘箱,烘炉
evaporate [i'væpəreit] 蒸发
energy efficient 高能效的,节能的

【Key Sentences】

1. Polymer materials are mostly **thermoplastic**, that is, When the polymer is heated to a certain temperature, it softens and melts, then becomes the **viscous fluid** with a certain **fluidity**, which will re-solid and turn into the solid after being cooled.

2. After the fiber web heated, part of the fibers or **hot-melt powders** soften and melt, causing the fibers adhere with each other, so the fiber web is reinforced and become thermal bonding nonwoven materials after being cooled.

Fig 6.25 shows the schematic of thermal calendar bonding. Fig 6.26 shows the area and point bonding.

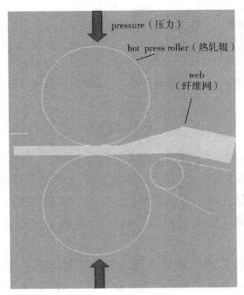

Fig 6.25 Schematic of thermal calendar bonding(热轧黏合)

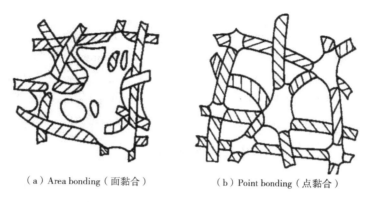

(a) Area bonding（面黏合）   (b) Point bonding（点黏合）

Fig 6.26   Area and point bonding（面黏合与点黏合）

3. Bonding is achieved using a **calendar process** of controlled hot air, **infrared heaters** or by high frequency or **ultrasonic welding**.

Fig 6.27 shows the shematic of ultrasonic bonding.

4. Thermal bonding requires less heat compared with the heat required to **evaporate** water from the binder, so it is more **energy efficient**.

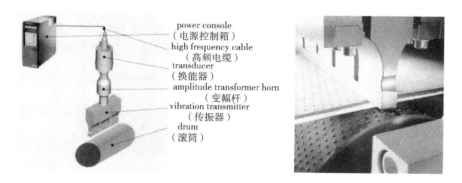

Fig 6.27   Shematic of ultrasonic bonding（超声波黏合示意图）

【Text】

热黏合常使用三种类型的纤维原料,每种纤维原料的适用范围不尽相同:纤维可以是相同类型的,且具有相同的熔点,如果在局部区域进行加热,这是令人满意的,但是如果使用整体黏合,则所有纤维可能熔化成具有很小或没有价值的塑料片[1];可使用可熔纤维与具有较高熔点的纤维或非热塑性纤维的混合物[2]。这在大多数条件下(热黏合效果)是令人满意的,除非可熔纤维完全熔化,失去其纤维性质并导致纤维网厚度坍塌[3];可熔纤维可以是双组分纤维,即挤出的纤维,其中高熔点聚合物芯被低熔点聚合物的皮包围[4]。这是一种理想的加工材料,因为纤维芯不会熔化,而是支撑纤维状态的护套[5]。

【Key Words】

fibrous ['faibrəs] 纤维构成的，纤维状的
melting point 熔点
localised 限局性的，局部的
integral ['intigrəl] 完整的，整体的
meltable 可熔的，易熔的

collapse [kə'læps] 倒塌，坍塌
bicomponent 双组分
extrude [ik'stru:d] （被）挤压出
sheath [ʃi:θ] 护层，护皮
ideal 完美的，理想的

【Key Sentences】

1. It is satisfactory that if the heat is applied at **localised** spots, but if **integral** bonding is used it is possible that all the fibers will melt into a small or no-value plastic sheet.

2. A mixture of **meltable** fibers and fibers with higher melting points or non-thermoplastic fibers can be used.

3. This is satisfactory in most conditions except where the fusible fiber melts completely, losing its fibrous nature and causing the web to **collapse** in thickness.

4. The fusible fiber may be the **bicomponent** fiber, that is, a fiber **extruded** with a **core** of high melting point polymer surrounded by a **sheath** of lower melting point polymer.

Fig 6.28 shows the structure of bicomponent fiber.

5. This is an **ideal** processing material because the core of the fiber does not melt, but rather supports the sheath in the fibrous state.

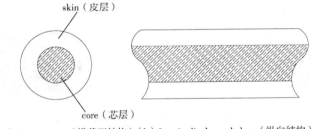

(a) Cross section（横截面结构） (b) Longitudinal morphology（纵向结构）

Fig 6.28　Structure of bicomponent fiber（双组分纤维的结构）

【Text】

热黏合也可以在无压力下进行加工。在这种情况下，纤维网通过热风烘箱，使可熔化的成分（热熔纤维或热熔粉末）融化，此方法用于生产高蓬松织物。产品（热风黏合）与喷洒黏合非织造材料类似，只是在这种情况下各处黏合均匀，对非织造织物的厚度几乎没有限制[2]。热黏合织物的用途与喷洒黏合非织造织物的用途基本相同。

热黏合也可以在高压下进行处理。纤维网在两个大的加热辊（热轧辊）之间经过，这两个热轧辊同时熔化可熔纤维并压缩纤维网[3]。如果纤维网单位面积质量不重，则加热过程非常迅

速,并且该过程可以高速(300m/min)进行[4]。虽然可以通过改变混合物中可熔纤维的百分比来调节黏合量,但产品往往是致密且重度黏合的[5]。热黏合非织造布典型的特性是高强度,有非常高的模量和刚度,但也有好的弯曲恢复性[6]。其主要用途是一些土工布,一些衣服和鞋子中的加强筋,一些过滤介质和屋顶膜。虽然热轧黏合非织造布非常坚固,但是在整个纤维网上黏合制成的织物(面黏合)对于许多用途而言太硬并且非纺织品状[7]。使用点黏合更为常见,其中一个热轧辊刻有一种图案,该图案将辊子之间的接触程度限制到大约5%的面积[8]。黏合被限制在压辊接触的那些点上,并且大约95%的纤维网未黏合[9]。黏合点的面积、形状和位置非常重要。由于未黏合区域(大),以这种方式制造的织物是柔韧的并且相对柔软,同时它们保持合理的强度。这些织物具有许多用途,例如,作为簇绒地毯的基材、土工织物、过滤介质、保护/一次性衣服、涂料基材以及覆盖物。

【Key Words】

high fluffy fabric　高度蓬松织物
heated rollers　加热辊
mass per unit area　单位面积的质量
dense [dens]密集的,稠密的
modulus　模量
stiffness [ˈstifnəs] 硬度
bending resilience [riˈziliəns] 弯曲弹性
geotextiles　土工布;土工织物
stiffener [ˈstifnə] 加强筋

filtration [filˈtreiʃn] 过滤
roofing [ˈruːfiŋ] 盖屋顶用的材料,盖屋顶
membrane [ˈmembrein] (可起防水、防风等作用的)膜状物
be engraved with　镌刻
roughly [ˈrʌfli] 大约,大致,差不多
substrate [ˈsʌbstreit] 底层,基底,基层
tufted carpets　簇绒地毯

【Key Sentences】

1. The fiber web pass through a hot air **oven** to cause the meltable component to melt.

2. The products are similar to spray-bonded nonwoven fabrics except that in this case the bonding is uniform throughout the fiber web and there is **virtually** no limit to the thickness of the nonwoven fabric made.

3. The fiber web pass between two large **heated rollers** (calender rollers), both of which melt the fusible fiber and compress the web at the same time.

4. If the fiber web is not too heavy in **mass per unit area**, the heating process is very fast and the process can be carried out at high speed (300 m/min).

5. Although the bonding amount can be adjusted by changing the percentage of meltable fibers in the mixture, the products are often **dense** and heavily bonded.

6. The typical characteristics of thermal bonding nonwovens are high strength, very high modulus and stiffness, but also good **bending resilience**.

7. Although it is very strong, the nonwoven fabric produced with bonding all over the web (area bonding) is too stiff and **non-textile-like** for many uses.

8. It is far more common to use point bonding, in which one of the calender rollers **is engraved**

with a pattern that limits the degree of contact between the rollers to roughly 5% of the total area.

Fig 6.29 shows the calender roller engraved with a pattern.

9. The bonding is limited to those rollers contact points and leaves **roughly** 95% of the fiber web unbonded.

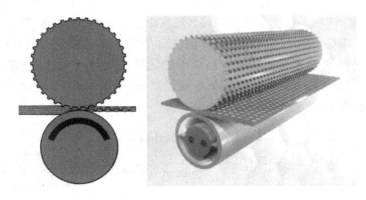

Fig 6.29　The calender roller engraved with a pattern（刻有花纹图案的热轧辊）

【Text】

针刺被广泛用作将纤维网黏合在一起的机械手段[1]，其基本原理：将带倒钩刺的刺针插入纤维网中，迫使一些纤维向下，使它们与纤维网下层的纤维缠结[2]；在刺针向上运动时能发生进一步的缠结。将特殊设计的针头和不同特征的纤维网组合在一起可以获得各种各样的特性；具有机织物和针织物特性的非织造布以及羊毛和丝绒的外观和性质可以通过针刺生产[3]。

针刺的基本原理很简单，其加工过程如下：纤维网通过压网辊式、压网帘式或双滚筒式三种方式输送后，被拖网板和剥网板引导[4]。在拖网板和剥网板之间，纤维网被大量的刺针穿透，刺针通常被制成三角形并且在三个边缘切入倒钩刺[5]。常用的针刺机类型有三种："下刺法"，因为它向下推动纤维；类似地，"上刺法"将纤维向上推；在尝试制作致密的纤维毡时，采用上刺和下刺相结合优点突出，而不是在同一方向上连续针刺[6]；出于这个原因，一些针刺机制成"双向刺法"，也就是一个针板向下刺，另一个针板向上刺[7]；虽然这种双向针刺机比单向针刺机更好，但它的缺点是在上刺开始之前完成所有的下刺[8]。

【Key Words】

needlepunching ['niːdl 'pʌntʃiŋ] 针刺；
　针刺法
barbed [bɑːbd] 有倒钩的
downward ['daunwəd] 下降的；向下的
upward ['ʌpwəd] 向上的；朝上的
fleece [fliːs] 羊毛状织物
velours　天鹅绒；丝绒；灯芯绒

bed plates　拖网板
stripper plates　剥网板
penetrate ['penətreit] 穿过；进入
triangular [traiˈæŋgjələ(r)] 三角的；三角形的
barbs [bɑibz] 倒钩，倒刺
needleloom　针刺机

Project Six Nonwoven Fabrics and Processing Technology(非织造物与非织造技术)

down-punch　下刺
up-punch　上刺
felt [felt]毛毡

double-punch　双刺
board　针板
unidirectional　单向的;单方面的

【Key Sentences】

1. **Needlepunching** is widely used as a means of mechanically bonding the fiber web together.

2. **Barbed** needles are inserted into the fiber web, which force some of the fibers **downward** to be entangled with fibers in under layers of the web.

Fig 6.30 shows the schematic of needlepunching.

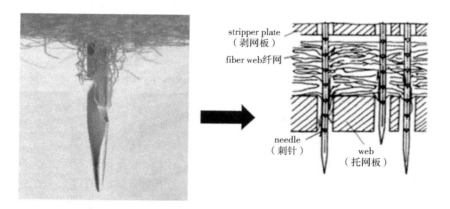

Fig 6.30　Schematic of needlepunching(针刺原理图)

3. Nonwoven fabrics with the characteristics of woven fabrics and knitted fabrics and the appearance and properties of **fleece** and **velours** can be produced by needlepunching.

4. The fiber web is guided by the bed and stripper plates after being conveyed by three ways, the **pressure-web roller** type, the **pressure-web curtain** type or the **double cylinder** type.

Fig 6.31 shows different conveying forms of fiber web for needlepunching.

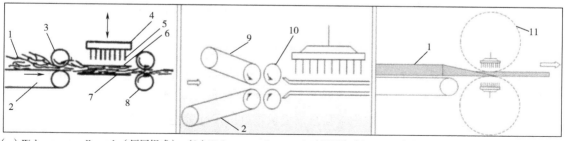

(a) Web pressure roller style (压网辊式)　(b) Web pressure lattice style (压网帘式)　(c) Two drum style (双滚筒式)

Fig 6.31　Conveying forms of fiber web for needlepunching(针刺纤维网输送形式)
1—web(纤网)　2—transport lattice(输送帘)　3—web pressureroll(压网辊)　4—needle plate(针板)
5—needle(刺针)　6—stripper plate(剥网板)　7—bed plate(托网板)　8—drawing roller(牵拉辊)
9—web pressure lattice(压网帘)　10—feeding roller(喂入辊)　11—drum(滚筒)

201

5. The fiber web between the **bed** and **stripper plates** is **penetrated** by a large number of needles which are usually made **triangular** and have **barbs** cut into the three edges.

Fig 6.32 shows the structure composition of needles.

6. There are some advantages in combining an **up-punch** with a **down-punch** when trying to make a dense **felt**, rather than to punch continually in the same direction.

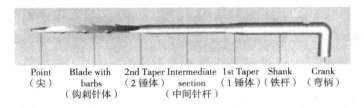

Point Blade with 2nd Taper Intermediate 1st Taper Shank Crank
（尖） barbs （2锤体） section （1锤体）（铁杆）（弯柄）
（钩刺针体） （中间针杆）

Fig 6.32 Structure composition of needles(刺针的结构组成)

7. For this reason, some **double-punch** needllooms were made, that is, one **board punch down** and the other board punch up.

Fig 6.33 shows different needlepuching types.

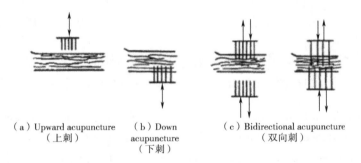

(a) Upward acupuncture　　(b) Down　　(c) Bidirectional acupuncture
　　　（上刺）　　　　　acupuncture　　　　　（双向刺）
　　　　　　　　　　　　（下刺）

Fig 6.33 Types of different needlepuching(不同针刺类型)

8. Although this type of double-punch needleloom is better than **unidirectional** punching, it has the disadvantage that all the down-punch is completed before the up-punch even starts.

【Text】

针刺非织造织物具有高断裂强度和高撕裂强度,但模量低且伸长恢复性差[1]。由于这两个原因,任何可能承受负荷的针刺非织造织物必须具有某种形式的增强,以控制伸长[2]。例如,针刺地毯可以用化学黏合剂浸渍,该化学黏合剂提供更好的尺寸稳定性并增加耐磨性。针刺非织造织物通常称为针刺毡,它们广泛用于气体过滤介质和一些湿过滤。与机织物相比,非织造物的主要优点是均匀的,因此,非织造过滤器的所有区域可用于过滤[3]。

针刺非织造织物用作土工织物,但鉴于其模量低,主要应用于去除水而不是增强[4]。例如,可以通过用土工织物作为沟槽衬里来形成简单的排水沟,土工布通过过滤细颗粒,使排水沟保持多年的畅通[5]。许多合成革制造商认为(针刺毡)结构应该与天然皮革相似。在这种情况下,合成皮革的衬底或基布是针刺毡[6]。这种针刺毡的生产方法是在强烈针刺后使热收缩的纤维收缩,以使毛毡更致密[7];收缩后,使用黏合剂浸渍毡,该黏合剂填充毡中的空隙但不黏附到纤维上,否则毡会变得非常硬;然后通过用一层或两层合适的聚合物涂覆表面来完成。针刺毡还广泛用于家庭和商业地毯;在许多情况下,地毯可以具有丝绒或毛圈表面以改善外观[8]。丝绒和毛圈都是在单独的针刺操作中使用特殊的针头和针织机中的特殊托网板生产的。

【Key Words】

breaking tenacity　断裂强度
tear strength　撕裂强度
be subjected to　受到,被……折磨
be impregnated［'impregneit］with　使饱和
dimensional［dai'menʃənl］空间的,尺寸的
resistance to wear　耐磨性
practically［'præktikli］几乎,差不多
homogeneous［ˌhɒmə'dʒi:niəs］相同的,同种的
in view of　鉴于,考虑到
drain　下水道,排水管

trench［trentʃ］沟,渠
line　(用…)做衬里,(在某物内部)形成一层
filter out　滤除,滤掉
backing　背衬
foundation　基布
heat-shrinkable［hi:t'ʃriŋkəbl］热缩的
binder［'baində(r)］黏合剂
voids［vɔidz］空白
stiff　硬的
velour［və'luə(r)］丝绒
loop pile　毛圈

【Key Sentences】

1. Needlepunchednonwoven fabrics have a high **breaking tenacity** and also a high **tear strength**, but its modulus is low and the recovery from extension is poor.

2. For these last two reasons, any needlepunching nonwoven fabric which is likely to **be subjected to** a load need have some reinforcement and can control the extension.

3. Theprincipal advantage is that the nonwoven is **practically homogeneous** in comparison with a woven fabric so that the whole area of a nonwoven filter can be used for filtration.

4. Needlepunched nonwovens are used in geotextiles, but **in view of** the low modulus, their application is mainly in removing water rather than as reinforcement.

5. For instance, a simple **drain** may be formed by **lining** a **trench** with a geotextile which will keep the drain open for many years by **filtering out** fine particles.

6. In this case, the needlefelt is used as **backing** or **foundation** of the synthetic leather.

7. This needle felt is produced by shrinking the **heat-shrinkable** fibers after a intense needling to make the felt even more dense.

8. In many cases, the carpet may have either a **velour** or a **loop pile** surface to improve the appearance.

【Text】

　　水刺法(水力纠缠)是利用高压水流来机械黏合纤维网的另一种方法[1]。它被发明为一种产生类似于针刺的缠结的方法,但使用较轻的网。水力缠结意味着这个过程取决于在非常高的压力下通过直径非常小的喷射孔工作的水流[2]。这种极细的射流很容易破碎成液滴,特别是当水通过孔有湍流时[3]。如果形成液滴,则射流中的能量仍将大致相同,但它将扩散到更大的纤维网区域,使得单位面积的能量更少[4]。因此,射流设计以避免湍流并产生针状水流是至关重要的[5]。喷嘴规则排列,且纤网由多孔筛网拖持在喷嘴下连续通过,该多孔筛网去除大部分水[6]。确切地说,喷射器下方的纤网发生什么尚不清楚,但很明显,水中的湍流撞击纤网后,纤维末端会被扭曲在一起或缠绕在一起[7]。众所周知,拖网帘对过程至关重要;在其他所有变量保持不变的情况下,更换拖网帘将彻底改变所生产的非织造织物[8]。

　　虽然与大多数黏合系统相比,(水刺)设备具有更高的生产量,但特别是与针刺相比,它们仍然非常昂贵且需要大量电力。另一个相当大的问题在于在正确的 pH 和温度下向喷射器供应清洁的水[9]。因需要大量的水,水必需要回收利用。但在完成水刺过程(后),水会吸收气泡、短小纤维、纤维润滑剂或纤维整理剂,所以,有必要在回收利用之前清除(水中)所有其他东西[10]。因此,水过滤过程比运行(水刺)其他设备更困难。水刺非织造布用途包括擦拭巾、外科医生服、一次性防护服和背衬织物。

【Key Words】

Hydroentanglement = spunlacing　水刺
orifices ['ɔrifisiz] 孔,穴,腔
liable ['laiəbl] 很可能的
break up into　破碎成
turbulence ['tə:bjələns]（空气和水的）
　湍流,涡流,紊流
roughly ['rʌfli] 大约,大致,差不多
spread　使展开,扩散
needle-like　针状
critical　关键的,至关紧要的
perforate ['pə:fəreIt] 打孔,穿孔,打眼
underneath [ˌʌndə'ni:θ] 在…底下
supporting screen (spunlacing sieve + supporting sieve)　拖持网,拖网帘
variable ['veəriəbl] 变量,可变因素
constant　不变的,固定的
profoundly [prə'faundli] 完全地,彻底地
considerable [kən'sidərəbl] 相当多的
lie in　在于
recycling [ˌri:'saikliŋ] 回收利用,再利用
lubricant ['lu:brikənt] 润滑剂
finish　整理剂
wipes [waips] 擦,揩,拭,(湿)抹布
surgeons gowns　外科医生长袍

【Key Sentences】

1. **Hydroentanglement** is another means of mechanically bonding webs by the use of high-pressure jets of water.

2. Hydroentanglement implies the process depends on jets of water working at very high pressures through jet **orifices** with very small diameters.

Fig 6.34 shows schematic diagram of spunlaced process.

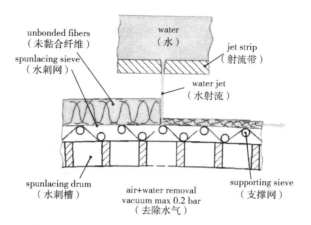

Fig 6.34 Schematic diagram of hydroentanglement process(水刺工艺原理图)

3. A very fine water jet is **liable** to **break up into** droplets, particularly if there is any **turbulence** in the water passing through the orifice.

4. If droplets are formed, the energy in the jet stream will still be **roughly** the same, but it will **spread** over a much larger area of web so that the energy per unit area will be much less.

5. Consequently, the design of the jet to avoid turbulence and to produce a needle-like stream of water is **critical**.

Fig 6.35 shows the spunlacing jet.

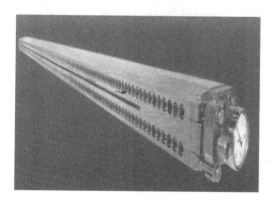 

Fig 6.35   Spunlacing jet(水刺头)

6. The jets are arranged regularly and the web is passed continuously under the jets held up by a **perforated screen** which removes most of the water.

Fig 6.36 shows different arrangement of spunlacing jets.

7. Exactly what happens to the web **underneath** the jets is not known, but it is clear that single fibers become twisted together or entangled by the turbulence in the water after it has hit the web.

8. Changing the **supporting screen** with all other **variables** remaining **constant** will completely

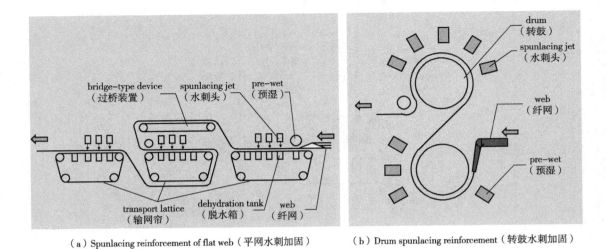

Fig 6.36　Different arrangement of spunlacing jets(水刺头排列形式)

alter the fabric produced.

Fig 6.37 shows different weave structure of the supporting screen.

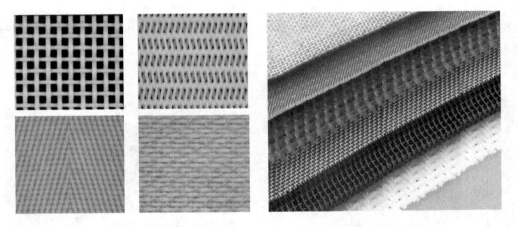

Fig 6.37　Weave structure of the supporting screen(拖网帘组织结构)

9. The other **considerable** problem **lies in** supplying clean water to the jets at the correct pH and temperature.

10. The water picks up air bubbles, bits of fiber and fiber **lubricant** or fiber **finish** in passing through the process, so it is necessary to remove everything before **recycling**.

Part 2　*Trying*

1. Talk about what are the specific requirements of thermal bonding reinforcement and spunlaced reinforcement for fiber properties (thermal properties, moisture absorption, etc.) and processing technology?

Project Six　Nonwoven Fabrics and Processing Technology(非织造物与非织造技术)

2. Compare and analyze production cost, product performance and environmental pollution about the three kinds of reinforcement methods, such as mechanical bonding, chemical bonding and thermal bonding.

	Mechanical bonding	Chemical bonding	Thermal bonding
Production cost			
Product performance			
Environmental pollution			

Part 3　*Thinking*

1. After the chemical bonding reinforcement, what are the characteristics of the nonwovens reinforced by the saturated impregnation method and the spraying bonding method? What kind of nonwoven products are they suitable for processing? Give 2 to 3 examples. Reason?

2. Spunlaced reinforcement is a very popular method of mechanical reinforcement, so is there any requirement for water in the reinforcement process? What do you think are the requirements?

Task Four　Nonwoven Technology of Polymer Extrusion (聚合物挤压技术)

【Text】

纺丝成网包括聚合物原料挤出长丝,拉伸长丝并将它们铺成网[1]。由于铺网和黏合通常是连续的,因此,该过程代表了在一步法中从聚合物到织物的最短可能的纺织路线[2]。除此之外,纺丝成网工艺已经变得更加多样化。当首次推出时,只有大型、非常昂贵的具有大量生产能力的机器可供使用,但现在已经开发了更小且相对便宜的机器,允许更小的非织造生产商使用纺丝成网工艺路线[3]。

进一步的发展使得在纺丝成网机上可以生产超细纤维,从而赋予其很多优点,如较好的长丝分布,因纤维之间较小孔隙而具有更好的过滤,更柔软的手感;也使制造更轻质织物成为可能[4]。由于这些原因,纺丝成网(非织造布)的增产速度比任何其他非织造工艺快。

【Key Words】

spun laying　纺丝成网
extrusion [ikˈstruːʒn]　挤压
polymer [ˈpɔlimə(r)]　聚合物
continuous [kənˈtinjuəs]　持续的;连续的
versatile [ˈvəːsətail]　多样化的,多功能的
inexpensive [ˌinikˈspensiv]　不昂贵的
route [ruːt]　(工艺)路线
microfiber　超细纤维
pore [pɔː(r)]　孔隙

【Key Sentences】

1. **Spun laying** includes **extrusion** of the filaments from the **polymer** raw material, stretching the filaments and laying them into a web.

Fig 6.38 shows the schematic diagram of spunlaying process.

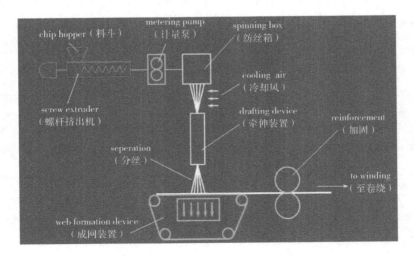

Fig 6.38　Schematic diagram of spunlaying process(纺丝成网工艺原理图)

2. Because laying and bonding are generally **continuous**, this process represents the shortest possible textile route from polymer to fabric in one step.

3. When first introduced, only large, very expensive machines with large production capabilities were **available**, but today, much smaller and relatively **inexpensive** machines have been developed, permitting the smaller nonwoven producers to use the spun laying **route**.

4. Further developments have made it possible to produce **microfibers** on spun laying machines giving the advantages of better filament distribution, smaller **pores** between the fibers for better filtration, softer feel and also the possibility of making lighter-weight fabrics.

【Text】

纺丝成网从(聚合物)挤压开始。聚酯和聚丙烯是目前最常见的,但也可以使用聚酰胺和聚乙烯。将聚合物切片连续加入螺杆挤出机中,该螺杆挤出机将液体聚合物输送到计量泵,然后输送到一组喷丝板,或者输送到矩形喷丝板,每个喷丝板含有非常多的喷丝孔[1]。液体聚合物被泵输送通过每个孔后迅速冷却至固态,但不可避免地会发生一些液态拉伸或牵伸[2]。到现阶段,该技术类似于合成纤维生产中的纤维或长丝挤出,除了速度更高外,但这里的技术有一些区别[3]。

在纺丝成网中,最常见的获得正确模量的牵伸长丝的方式是气流牵伸。气流牵伸中,高速气流吹过长丝沿长管向下运动,空气速度和管长条件需选定以便在长丝中产生足够的张力引起牵伸[4]。在某些情况下,气流拉伸是不够的,并且必须使用罗拉拉伸,就像正常的纺织品挤出中那样,但是罗拉牵伸更复杂并且趋于减慢生产过程,因此,优选空气拉伸[5]。

【Key Words】

polyethylene [ˌpɔli'eθəli:n] 聚乙烯	矩形的
chips [tʃips] 碎片	inevitably [in'evitəbli] 不可避免地
screw extruder 螺杆挤压机	take place 发生
metering pump 计量泵	except that 除了…
a bank of 一组	air drawing 气流牵伸
spinnerette 喷丝板	roller drawing 罗拉牵伸
rectangular [rek'tæŋgjələ(r)] 长方形的,	velocity [və'lɔsəti] 速度,高速,快速

【Key Sentences】

1. The polymer **chips** are fed continuously to a **screw extruder** which delivers the liquid polymer to a **metering pump** and then to a bank of **spinnerettes**, or alternatively to a **rectangular** spinnerette plate, each containing a very large number of holes.

Fig 6.39 shows the polymer chips, Fig 6.40 shows the screw extruder, Fig 6.41 shows the spinnerette.

2. The liquid polymer pumped through each hole is cooled rapidly to the solid state but some stretching or drawing in the liquid state will **inevitably take place**.

Fig 6.39　Polymer chips(聚合物切片)

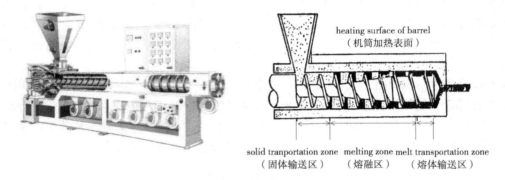

Fig 6.40　Screw extruder(螺杆挤出机)

Fig 6.41　Spinnerette(喷丝板)

3. Up to this stage, the technology is similar to the fiber or filament extrusion in synthetic fiber production, **except that** the speeds are higher, but here the technologies tend to divide.

4. Inspun laying, the most common form of drawing the filaments to obtain the correct modulus is

air drawing, in which a high **velocity** air stream is blown past the filaments moving down a long tube, the conditions of air velocity and tube length being chosen so that sufficient tension is developed in the filaments to cause drawing.

5. In some cases, air drawing is not adequate and roller drawing has to be used as in normal textile extrusion, but roller drawing is more complex and tends to slow the process, so that air drawing is **preferred**.

【Text】

熔喷工艺是另一种以高生产率生产极细纤维的方法[1]。当聚合物离开挤出孔时,它会受到达到或高于聚合物熔点的高速热空气流的冲击(喷吹),这会破坏流体并拉伸许多长丝直到很细[2]。在后一步骤,冷空气与热空气混合并使聚合物固化。有时候,长丝会破碎成短纤维,很可能在聚合物仍处于液态时发生这种情况,因为如果在后面发生这种情况,这将意味着固体纤维被施加了高张力,这会导致(长丝)在断裂之前(形成)拉伸[3]。在气流成网和纺丝成网中,以这种方式生产的细短纤维被收集成网在传送带上。与纺丝成网法相比,最大的区别在于熔喷纤维非常细,因此,纤维与纤维之间的接触更多,并且纤维网具有更高的完整性[4]。

不使用任何(其他)形式的黏合适合许多最终用途,该材料仅仅是松散纤维网[5]。这些用途包括用于空调和个人面罩的超细过滤器、溢油吸收剂和个人卫生用品。在其他情况下,熔喷纤维网可以层压到另一种非织造材料上,特别是纺丝成网或熔喷纤维网本身可以黏合,但必须仔细选择(层压)方法以避免破坏超细纤维的开松(特性)[6]。直接黏合或层压形式的非织造织物可用于医院、农业和工业中的透气防护服,作为具有良好绝缘性能的电池隔膜、工业擦拭物和服装中间层[7]。

当然,纺丝成网和熔喷成网后的非织造布也可以通过针刺、水刺、化学黏合等其他形式的加固方法进行二次加固,这主要取决于产品的最终用途、产品外观形态等要求。

【Key Words】

melt-blown 熔喷
break up 使破碎,使破裂
solidifies [sə'lidifaiz] (使)凝固,变硬
conveyor [kən'veiə(r)] 传送带
integrity [in'tegrəti] 完整,完好
ultrafine [ˌʌltrə'fain] 超细的
air conditioning 空调调节系统
face mask [mɑ:sk] 面罩
oil-spill absorbents 溢油吸收剂
hygiene ['haidʒi:n] 卫生

laminate ['læminət] 把…制成薄板(层压)
spoil [spɔil] 破坏,搞坏
openness 开松
breathable protective clothing 透气防护服
agriculture ['ægrikʌltʃə(r)] 农业
battery separator 电池隔膜,电池隔板
clothing inter-lining 服装衬里

【Key Sentences】

1. The process of **melt-blown** is another method of producing very fine fibers at high production

rates.

2. When the polymer leaves the extrusion holes, it is hit by a high speed stream of hot air at or even above its melting point, which **breaks up** the flow and stretches the many filaments until they are very fine.

Fig 6.42 shows the schematic diagram of melting-blown process.

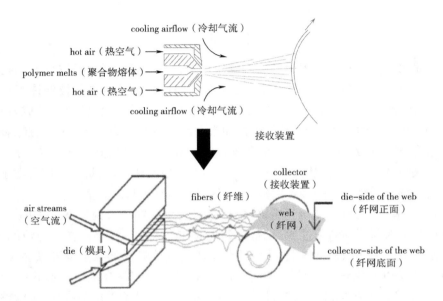

Fig 6.42 Schematic diagram of melting-blown process(熔喷工艺原理示意图)

3. At some point, the filaments break into staple fibers, but it seems likely that this happens while the polymer is still liquid, because if it happened later, this would imply that a high tension had been applied to the solid fiber, which would have caused drawing before breaking.

4. The big difference is that in melt blowing the fibers are extremely fine so that there are many more fiber-to-fiber contacts and the web has greater **integrity**.

5. No form of bonding is used for many end-uses and the material is simply a web of loose fibers.

6. In other cases, the melt-blown web may be **laminated** to another nonwoven, especially a spun-laid one or the melt-blown web itself may be bonded but the method must be chosen carefully to avoid **spoiling** the **openness** of the very fine fibers.

7. Bonded or laminated nonwoven fabrics can be used for **breathable protective clothing** in hospitals, **agriculture** and industry, as **battery separators**, industrial wipes and **clothing inter-linings** with good insulation properties.

Part 2 *Trying*

1. According to the characteristics of the processing technology, compare the difference between the spun laying nonwovens and melt blown nonwovens on the net structure and performance.

Project Six Nonwoven Fabrics and Processing Technology(非织造物与非织造技术)

	Net structure	Performance
Spun laying		
Melt-blown		

2. Taking the process of pure polypropylene staple fiber yarn woven fabric, polypropylene staple fiber needle-punched nonwovens, polypropylene staple fiber melt-blown nonwovens as examples, write their production process.

a. Production process for pure polypropylene staple fiber yarn woven fabric: _____

b. Production process for polypropylene staple fiber needle-punched nonwoven: _____

c. Production process for polypropylene staple fiber melt-blown nonwovens: _____

Part 3 *Thinking*

1. Is the spun laying technology or melt-blown technology related to the manufacture of chemical fibers? What are the clear requirements for fibers?

2. What are the advantages of nonwoven technology compared to woven technology and knitting technology in terms of process flow or economic cost?

Project Seven　Textile Dyeing and Finishing（纺织品染整）

Task One　Pretreatment Technology（前处理技术）

【Text】

大多数纺织材料和织物在染色和整理之前需要进行预处理。这种预处理必要的原因是确保染色、印花和整理是可接受的、可预测的和可再现的[1]。无论是天然的,还是人工引入的现存杂质都必须除去[2]。这些预处理对所有的布料基本相同,但它们在一定程度上随纤维类型而变化[3]。预处理工序主要包括烧毛、退浆、煮练、丝绸脱胶、漂白和偶尔的棉丝光处理。

烧毛是染色或印花所需的准备过程的第一个操作。烧毛去除伸出的纤维头端,以产生清晰、光滑和均匀的表面[4]。烧毛通常是将织物高速通过燃烧的气体火焰,然后在水中或退浆浴中骤冷以熄灭任何火花来完成的[5]。有时在煮练之后进行烧毛,因为在烧毛中,加热织物会增加从织物上去除浆料和尘土的难度。烧毛和退浆经常结合在一起,退浆与烧毛相结合可防止烧毛火焰或加热板对织物造成可能的潜在损伤[6]。

退浆可以去除织物上的浆料。在纱线织成面料之前,将浆料施加到纱线上,特别是经纱上;它们在纱线上形成保护涂层,并在织造过程中防止纱线摩擦或断裂[7]。在织物可以令人满意地染色之前首先可通过退浆来去除大部分浆料。但(退浆后)残留浆料阻碍纱线和纤维快速润湿,也会影响批量染色或连续染色中染料吸收[8]。因此,残留的浆料可通过煮练进一步去除。

【Key Words】

predictable [pri'diktəbl] 可预见的;可预料的;意料之中的
reproducible 可复制的
impurities [im'pjuəritiz] 杂质
present ['preznt] 现存的
approximately [ə'prɔksimətli] 大概,大约
singeing ['sindʒiŋ] 烧毛
desizing 退浆
scouring ['skauəriŋ] 煮练
degumming 脱胶
bleaching ['bliːtʃiŋ] 漂白
occasionally [ə'keiʒnəli] 偶然,偶尔

mercerization 丝光
preparatory [pri'pærətri] 预备的,筹备的
protruding [prə'truːdiŋ] 突出,伸出
quench [kwentʃ] 扑灭,熄灭
extinguish [ik'stiŋgwiʃ] 熄灭,扑灭
spark 火花
latent 潜在的
chafe [tʃeif] 摩擦
satisfactorily [ˌsætis'fæktərəli] 满意地,可靠地
residual [ri'zidjuəl] 剩余的,残留的

Project Seven Textile Dyeing and Finishing(纺织品染整)

【Key Sentences】

1. The reason why such preliminary treatments are necessary is to make sure that the dyeing, printing and finishing are acceptable, **predictable** and **reproducible**.

2. The **impurities present**, either natural or man-introduced, must be removed.

Fig 7.1 shows grey cloth and bleaching cloth.

3. These preliminary treatments are **approximately** the same for all cloths, but they vary to some extent with fiber type.

4. **Singeing** removes **protruding** ends of fibers in order to produce a clear, smooth, and uniform surface.

5. Singing is usually done by passing the fabric through a burning gas flame at high speed followed by **quenching** in water or in the desizing bath to **extinguish** any **sparks**.

6. When combined with singeing, desizing prevents possible **latent** damage to fabric from the singeing flames or heated plates.

7. Size form a protective coating over the yarns and keep them from **chafing** or breaking during weaving.

8. **Residual** size prevents the yarns and fibers from wetting quickly and can affect dye absorption in either **batch** or continuous dyeing.

(a) Grey cloth (坯布)

(b) Bleaching cloth (漂白布)

Fig 7.1 Grey cloth and bleaching cloth(坯布和漂白布)

【Text】

煮练是将纱线加工过程中可能黏附在纤维上的任何果胶、灰分、尘土、蜡或其他物质(如残留的浆料)减少到不会严重干扰接下来染色的操作过程[1]。在煮漂连续系统中,漂白在煮练后面,漂白是完成去杂的最后工序;漂白同时确保织物产生较好的白度并增加织物在染色或印花时均匀吸收染料的能力[2]。

具有碱性助洗剂的肥皂或合成洗涤剂是常见的精练剂[3]。对于蛋白质纤维,通常使用中性或微酸性合成洗涤剂。尽管(煮练加工)使用了化学品,但纤维中不会发生分子变化。在煮练中,去除在退浆过程中仍然残留在织物中的化学品[4]。漂白使用的特定化学品取决于纺织纤维;

纤维素纤维,如棉、亚麻和各种再生棉,可用氯化合物(次氯酸钠是所用化学品之一)或过氧化氢漂白[5]。过硼酸盐漂白剂可供家庭使用,但它们很少在商业上使用,因为过氧化氢(与过硼酸盐相比)以相同的方式(条件)反应并且效率更高[6]。

【Key Words】

pectin ['pektin] 果胶
ash [æʃ] 灰,灰烬
dirt [dəit] 尘土
wax [wæks] 蜡质
interfere [ˌintə'fiə(r)] with 干扰,阻碍
subsequent ['sʌbsikwənt] 随后的,接后的
purfication [ˌpjuərifi'keiʃ(ə)n] 提纯,可译为去杂
dyestuff ['daistʌf] 染料,染色剂
soap [səup] 肥皂
synthetic detergent [di'tə:dʒənt] 合成洗涤剂
alkali builders 碱性助洗剂
scouring agent 精练剂
acidic [ə'sidik] 酸性的
molecular [mə'lekjələ(r)] 分子的
chlorine ['klɔ:ri:n] compounds 氯化物
sodium ['səudiəm] hypochlorite [haipəu'klɔ:rait] 次氯酸钠
hydrogen ['haidrədʒ(ə)n] peroxide [pə'rɔksaid] 过氧化氢
perborate [pə'bɔ:reit] bleaches 过硼酸盐漂白剂

【Key Sentences】

1. Scouring is an operation in which any **pectin**, **ash**, **dirt**, **wax** or other substances that may have adhered to the fibers in the processing of the yarns is reduced to an amount which will not seriously **interfere with subsequent** dyeing.

2. At the same time, bleaching ensures that the fabric produces a good **whiteness** and increases the fabric's ability to absorb the **dyestuffs** uniformly when it is to be dyed or printed.

3. **Soaps** or **synthetic detergents** with **alkali builders** are common **scouring agents**.

4. During the scouring operation, chemicals applied during desizing that have remained in the fabric are removed.

5. Cellulosic fibers, such as cotton, linen and the various rayons, can be bleached with **chlorine compounds**(**sodium hypochlorite** is one of the chemicals used) or **hydrogen peroxide**.

6. **Perborate bleaches** are available for home use, but they are seldom employed commercially since hydrogen peroxide reacts in much the same way and is more effective.

【Text】

织物可以绳状或开幅形式漂白。在绳状漂白中,织物被挤在一起形成一个近圆形的物质,其足够松散以便渗透并且类似于大绳索;在开幅形式中,织物在张力下平整且光滑[1]。用于绳状棉或棉混纺织物练漂的典型步骤如下:织物依次喂入预洗机、浸碱箱、J型箱、水洗、过氧化氢容器、J型箱、水洗,最后进入储存箱[2]。开幅漂白在一些细微方面有所不同:织物在张力作用下被处理成开幅平整的形式,以防止形成折痕[3];织物通过带有一系列滚筒的蒸箱,而不是穿过J形

箱[4];织物以 80~120 码/min(73~110m/min)的速度穿过装置。

【Key Sentences】

open-width form　开幅形式
somewhat ['sʌmwɔt]　稍微,优点
mass　块状物
penetration [peni'treiʃ(ə)n]　渗透,穿透
resemble [ri'zemb(ə)l]　与…相似,像
typical ['tipik(ə)l]　典型的

prewasher　预洗机
caustic saturator　浸碱箱,煮练浴
peroxide saturator　漂白浴
storage bin　储存箱
crease [kri:s] mark　褶痕,皱痕
steamer ['sti:mə(r)]　蒸箱

【Key Sentences】

1. In rope bleaching, the fabric is pulled together to form a **somewhat** circular **mass**, which is loose enough for **penetration** and **resembles** a large rope.

2. **Typical** steps in scouring and bleaching for either cotton or cotton blend fabrics in rope form are as follows: Fabric is fed to **prewasher**, **caustic saturator**, J-box, washers, **peroxide saturator**, J-box, washers, and finally to **storage bins**.

3. The fabric is handled in open **flat** width under tension to prevent formation of **crease marks**.

4. Instead of passing through J-boxes, the fabric passes through **steamers** with a series of rollers.

【Text】

棉纤维用浓烧碱溶液浸透后,发生不可逆的剧烈溶胀,纤维横断面由扁平形转变为圆形,纵向的天然扭曲消失,长度缩短[1]。如果在对纤维施加一定的张力时浸浓碱,不使纤维收缩,纤维变为十分光滑的圆柱体,纤维表面对光线有规则地反射而呈现出光泽[2]。如果持续张力条件下水洗去除纤维上碱液,就基本上可以把棉纤维溶胀时的形态保留下来,且获得的光泽较耐久[3]。因此,棉织物在经纬向都施加张力条件下用浓碱浸渍,并经冲洗去碱后,织物不再收缩,可使织物获得如丝织物般的光泽[4],这就是丝光处理。丝光是一种用于纤维素纤维特别是棉纤维的化学整理,一般含有棉纤维的织物大都经过丝光处理。

丝光的步骤为用水浸润织物,用碱浸渍,保证丝光反应时间,拉幅成设定的织物尺寸,洗涤、中和、洗涤和干燥[5]。棉及其混纺织物的丝光处理除了改善光泽外,还提高化学活泼性,对染料亲合力增加,织物尺寸稳定性提高,强力、延伸性等都有所增加[6]。因此,丝光棉织物产品成本较高,终端消费品一般是高档 POLO 衫、T 恤、衬衫和商务袜。

【Key Words】

saturate ['sætʃəreit]　使浸透,渗透
concentrated caustic ['kɔ:stik]　浓碱
irreversibly　不可逆地
swell [swel]　膨胀,溶胀
shiny ['ʃaini]　有光泽的

durable gloss　持久的光泽
warp direction　经向
weft direction　纬向
silky sheen　丝绸般的光泽
mercerization　丝光处理

wet out 使…浸湿
tentering 拉幅
neutralizing ['nju:trəlaiziŋ] 中和
viability [ˌvaiə'biliti] 活性
affinity [ə'finəti] 亲和力
dyestuff ['daistʌf] 染料

dimensional stability 尺寸稳定性
extension [ik'stenʃn] 延伸性
ritzy ['ritsi] 高档的,奢华的
T-shirt T恤衫
U-shirt 衬衫
V-business sock 商务袜

【Key Sentences】

1. After the cotton fiber is **saturated** with **concentrated caustic**, the cotton fiber is **irreversibly swelled**, the cross-section of fiber is changed from flat to round, the **longitudinal natural twist** disappears, and the length is shortened.

Fig 7.2 shows the comparison of morphological structure of cotton fiber before and after mercerizing.

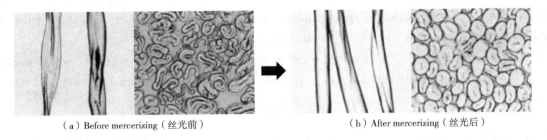

(a) Before mercerizing (丝光前)　　　　(b) After mercerizing (丝光后)

Fig 7.2　Comparison of morphological structure of cotton fiber before and after mercerizing
(丝光前后棉纤维形态结构对比图)

2. If the fiber is saturated with caustic under certain tension, the fiber will not shrink and becomes a very smooth cylinder, so the fiber surface reflects the light regularly and appears **shiny**.

Fig 7.3 Diagram of basic steps in mercerization.

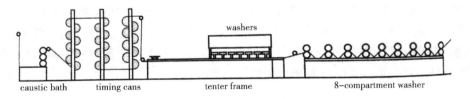

Fig 7.3　Diagram of basic steps in mercerization(丝光基本过程原理图)

3. If the water washing removes the caustic on the fiber under the continuous tension, it can basically preserve the swelled form of the cotton fiber and obtain the **durable gloss**.

4. Therefore, cotton fabric was saturated with concentrated caustic under the condition of tension applied at both **warp** and **weft direction** and then washed to remove caustic, so that the fabric can get the **silky sheen**.

5. Steps in mercerizing are **wetting out** the fabric with water; saturating with caustic; timing to permit mercerizing action; **tentering** to **set** fabric dimensions; washing; **neutralizing**; washing and drying.

6. In addition to improve gloss, the **mercerization** of cotton and its blended fabrics produce them with improved chemical **viability**, increased **affinity** for **dyestuffs**, enhanced fabric **dimensional stability**, increased strength and **extension**.

Fig 7.4 shows mercerized cotton products.

(a) T-shirt（T恤衫）

(b) Shirt（衬衫）

Fig 7.4　Mercerized cotton products(丝光棉产品)

Part 2　*Trying*

1. According to the previous learned knowledge, what kind of fabric need to do singeing treatment? Why?

_____

_____

_____

2. Go to the printing and dyeing factory in your city and buy the pure cotton fabric respectively

through desizing, scouring, bleaching and mercerization, analyze and compare their appearance, luster and handle.

	Appearance	Luster	Handle
Desizing			
Scouring			
Bleaching			
Mercerization			

Part 3 *Thinking*

1. Do all cotton fabrics need to go through the mercerizing process? Why?

_____
_____
_____

2. In fact, the pretreatment process is a chemical process? What effects do they have on environmental pollution?

_____
_____
_____
_____

Task Two Dyeing and Printing Technology(染色与印花技术)

【Text】

纺织品的颜色通常通过将着色剂施加到纺织品基材上而获得[1]。有两种方法可以为纺织品基材添加颜色：染色和印花。印花在基材局部增加颜色,而染色用颜色完全覆盖基材,并且通常具有在整个基材上获得均匀颜色分布的意图[2]。用于染色的着色剂可分为染料或颜料。

Project Seven Textile Dyeing and Finishing(纺织品染整)

【Key Words】

pleasure 使高兴,使愉快
derive from 由…起源
impart [im'pɑːt] 把(某性质)赋予
civilization [ˌsivəlai'zeiʃn] 文明,社会文明
hazardous ['hæzədəs] 危险的,有害的

dull [dʌl] 枯燥无味的,无聊的
colorant ['kʌlərənt] 着色剂,染(颜)料
substrate ['sʌbstreit] 基底,基层
locally 局部地
intension 意图,目的
pigment ['pigmənt] 色素,颜料

【Key Sentences】

1. The color of textiles is normally obtained by applying a **colorant** to the textile **substrate**. Fig 7.5 shows the dying process of fabrics.

Fig 7.5 Dying process of fabrics(织物的染色加工)

2. Printing adds color to the substrate **locally**; whereas dyeing completely covers the substrate with color, and usually with the **intension** of obtaining an even color distribution throughout the substrate.

Fig 7.6 shows different colors of dying fabrics. Fig 7.7 shows different printing patterns of fabrics.

Fig 7.6 Different colors of dying fabrics(不同颜色的染色织物)

221

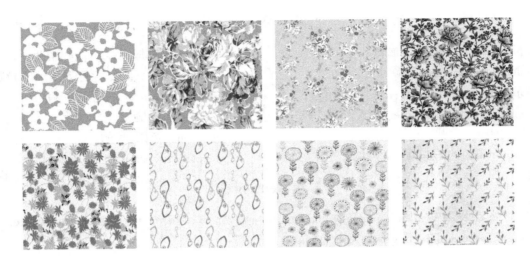

Fig 7.7　Different printing patterns of fabrics(不同印花图案织物)

【Text】

染料是水溶性的,且对纤维具有本质上的亲和力并且可以被吸收到纤维中[1]。颜料不是水溶性的,且对任何特定的纤维类型都没有特定的吸引力[2]。颜料通常黏附在纤维表面上。染料在纺织染色过程中比颜料使用得更广泛、更频繁[3]。

纺织品可以在从纤维到织物或某些服装的任何发展阶段通过以下方法染色[4]:散纤维染色(在纤维阶段);纱线染色(纤维被纺制成纱线之后);匹染(纱线被制成织物之后[5]);溶液染色或纺前染色(原液着色)[在再生纤维从喷丝板被挤压出去之前(湿法纺丝而成)][6];成衣染色(在一些品种的服装加工后)。

【Key Words】

water-soluble　水溶性的
substantively ['sʌbstəntivli]　实质上;实质性地
attraction　吸引力
stock dyeing　散纤维染色
yarn dyeing　纱线染色
piece dyeing　匹染
solution/dope dyeing　溶液染色
garment dyeing　成衣染色

【Key Sentences】

1. **Water-soluble** dyes have **substantively** affinity for fibers and can be absorbed into the fibers. Fig 7.8 shows the preparation of water-soluable dye solution.

2. Pigments are not water-soluble and possess no specific **attraction** for any particular fiber type.

3. Dyes are used far more widely and frequently than pigments in the textile dyeing process.

4. Textiles may be dyed at any stage of their development from fiber into fabric or certain gar-

(a) Natural environment-friendly dye (天然环保染料) (b) Dye solution (染液) (c) Reactive dyes (活性染色染料)

Fig 7.8 Preparation of water-soluable dye solution(可溶性染液的制备)

ments by the following methods.

Fig 7.9 shows stock dyeing, Fig 7.10 shows piece dyeing, Fig 7.11 shows bobbin yarn dyeing, Fig 7.12 shows garment dyeing.

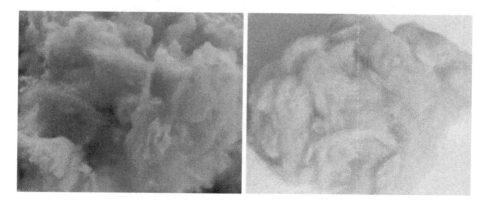

Fig 7.9 Stock dyeing(散纤维染色)

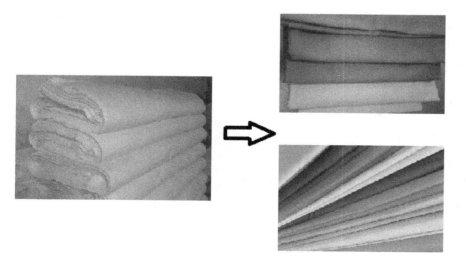

Fig 7.10 Piece dyeing(匹染)

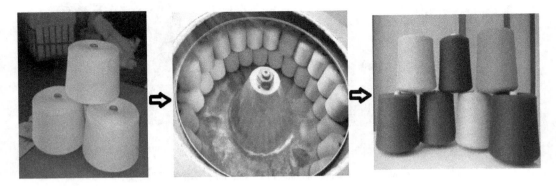

Fig 7.11　Bobbin yarn dyeing(筒纱染色)

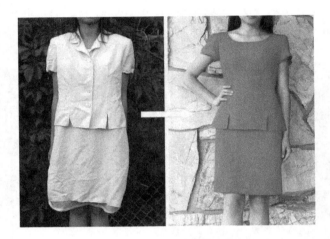

Fig 7.12　Garment dyeing(成衣染色)

5. **Piece Dyeing**, after the yarn has been constructed into fabric.

6. **Solution or dope dyeing**, before a man-made fiber is extruded through the spinnerette.

Fig 7.13 shows the composition of dyeing polymer solution/dope, Fig 7.14 shows the wet spinning of dyeing plymer solution/dope.

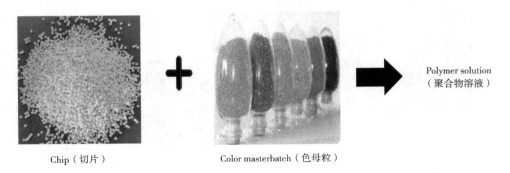

Fig 7.13　Composition of dyeing polymer solution/dope ［染色原液(纺丝液)的组成］

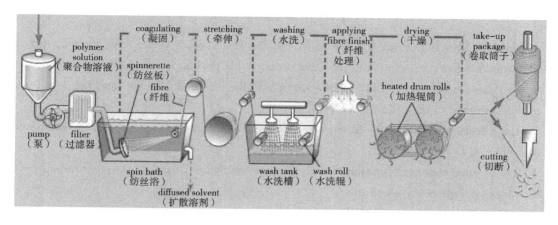

Fig 7.14 Wet spinning of dyeing plymer solution/dope(染色原液的湿法纺丝)

【Text】

　　根据染料性质及应用方法,常用的染料主要有以下几种:直接染料、酸性染料和媒染剂或金属络合染料、不溶性偶氮染料、阳离子染料、还原染料、硫化染料、活性染料、分散染料。

　　(染料)染色的基本过程大致可以分为三个阶段:当纤维投入染浴以后,染料先扩散到纤维表面,然后渐渐地由溶液转移到纤维表面,这个过程称为**吸附**[1];吸附在纤维表面的染料向纤维内部**扩散**,直到纤维各部分的染料浓度趋向一致[2]。染料在纤维上的扩散对染色速率及染色的匀染性起着决定性的作用[3];**固着**是染料与纤维结合的过程;染料和纤维不同,其结合方式也各不相同。一般来说,染料固着在纤维上存在着两种类型:一是纯化学性固着,指染料与纤维发生化学反应,而使染料固着在纤维上[4];二是物理化学性固着,由于染料与纤维之间的相互吸引(分子吸引力)及氢键的形成,而使染料固着在纤维上[5]。

【Key Words】

substantive/direct dye　直接染料
acid dyes　酸性染料
mordant/metallized dyes　媒染剂/金属络合染料
pigment　颜料
azoic/naphthol dyes　不溶性偶氮染料
cationic dyes　阳离子染料
vat dyes　还原染料
sulfur dyes　硫化染料
reactive dyes　活性染料

disperse dyes　分散染料
dye bath　染浴
adsorption　吸附
spread　(动词)扩散
uniform　均匀的,一致的
diffusion [diˈfjuːʒən]　扩散
play a role in　起作用
decisive [diˈsaisiv]　决定性的,关键的
pure chemical fixation　纯化学性固着
phy-chemical fixation　物理性化学固着

【Key Sentences】

1. When the fiber is put into the dye bath, the dye first spreads to the fiber surface, and then gradually transfers from the solution to the fiber surface, this process is called **adsorption**.

2. The dye adsorbed on the fiber surface **spreads** to the inside of the fiber until the dyes in each part of the fiber tend to be **uniform**.

3. The **diffusion** of dye on the fiber plays a **decisive** role in the dyeing rate and the homogenization of dyeing.

4. One is **pure chemical fixation**, which refers to the dyestuff fixed on the fiber because of the chemical reaction between dye and fiber;

5. Second is the **phy-chemical fixation**, which refers to the dyestuff fixed on the fiber because of the **mutual** attraction and the formation of hydrogen bonds between dye and fiber.

【Text】

印花使纺织材料形成局部着色,并经常用在织物或服装上制作彩色图案[1]。在印花过程中采用的每种颜色都必须在印花机器中的单独使用或特定位置应用[2];印花按照工艺来分,可分为直接印花、拔染印花和防染印花。直接印花是将含有染料的色浆直接印在白布或浅色布上,印有色浆处染料上染,获得各种花纹图案[3];拔染印花的过程是先染色后印花[4],印花色浆中含有能够破坏地色染料的化学药品(拔染剂)[5];拔染印花有拔白(花纹颜色为白色)和色拔(花纹有颜色)两种效果[6];防染印花的过程是先印花后染色,印花浆中含有能够防止地色染料上染的化学药品(防染剂)[7];防染印花有防白(花纹颜色为白色)和色防(花纹有颜色)两种效果。

印花方法也可以根据制作图案的印花设备进行分类[8]。筛网印花和滚筒印花是纺织品中最常用的两种印花方法[9];此外,纺织品印花中也采用喷墨印花、热转移印花和凸纹印花。

【Key Words】

localized coloration	局部染色	color discharge	色拔
direct printing	直接印花	white resist	防白
discharge printing	拔染印花	color resist	色防
restist printing	防染印花	screen printing	筛网印花
color paste	色浆	roller printing	滚筒印花
light colored cloth	浅色布	ink jet printing	喷墨印花
discharging agent	拔染剂	heat transfer printing	热转移印花
ground dyestuff	地色染料	reliefs	凸纹
white discharge	拔白		

【Key Sentences】

1. Printing produces **localized coloration** of textile materials and is often used to produce colored patterns on fabrics or garments.

2. Each color applied in a printing process must be applied in a separate step or position in the printing machine.

3. In **direct printing**, the **color paste** containing dyes are directly printed on the white cloth or **light colored cloth**, which is dyed by the dyes in color paste and obtain a variety of patterns.

Fig 7.15 shows the diagram of direct printing.

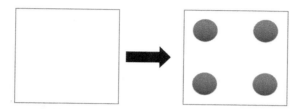

Fig 7.15   Diagram of direct printing(直接印花原理图)

4. The process of **discharge printing** is first dyeing and then printing.

5. The printing paste contains chemicals (**discharging agent**) which can destroy the **ground dyestuff**.

6. **Discharge printing** has two effects of **white discharge** (white pattern) and **color discharge** (colored pattern).

Fig 7.16 shows the diagram of discharge printing.

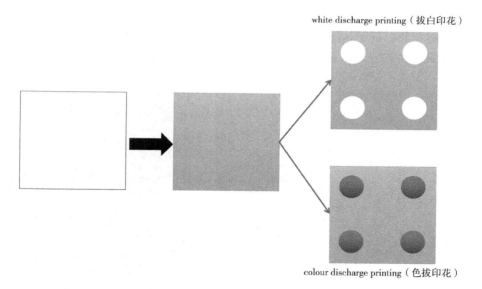

Fig 7.16   Diagram of discharge printing(拔染印花原理图)

7. The printing paste contains chemicals (resisting agent) that can prevent ground dyestuff dying on the cloth.

Fig 7.17 shows the diagram of resist printing.

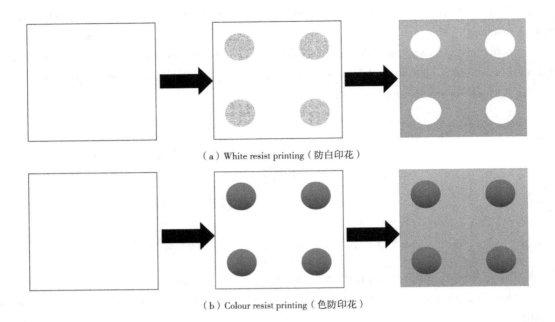

(a) White resist printing（防白印花）

(b) Colour resist printing（色防印花）

Fig 7.17　Diagram of resist printing（防染印花原理图）

8. Printing methods may also be classified according to the process used to produce the pattern.

9. **Screen printing** and **roller printing** are the two most common printing methods used in textiles.

Fig 7.18 shows the roller printing, Fig 7.19 shows the heat transfer printing, Fig7.20 shows the flat screen printing, Fig 7.21 shows the rotary screen printing, Fig 7.22 shows the ink jet printing.

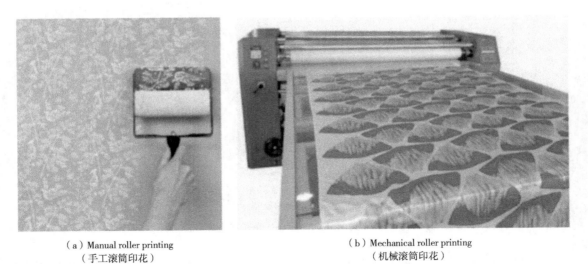

(a) Manual roller printing
（手工滚筒印花）

(b) Mechanical roller printing
（机械滚筒印花）

Fig 7.18　Roller printing（滚筒印花）

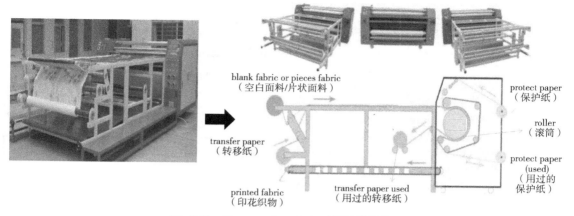

Fig 7.19　Heat transfer printing(热转移印花)

Fig 7.20　Flat screen printing(平网印花)

Fig 7.21　Rotary screen printing(圆网印花)

Fig 7.22　Ink jet printing(喷墨印花)

Part 2 *Trying*

1. Consult the relevant literature and point out that the following dyes are used for which kind of fibers or fabrics respectively: substantive or direct dyes, acid dyes and mordant or metallized dyes, azoic or naphthol dyes, cationic or basic dyes, vat dyes, sulfur dyes, reactive dyes, disperse dyes and pigments.

Dyes	Direct dyes	Reactive dyes	Disperse dyes
Suitable fibers or fabrics			

Dyes	Vat dyes	Acid dyes	Mordant or metallized dyes
Suitable fibers or fabrics			

Dyes	Azoic or naphthol dyes	Cationic or basic dyes	Sulfur dyes
Suitable fibers or fabrics			

2. Through careful observation for the textile and garment products in your daily life, which use more dyeing technology? Which use more printing technology?

_____

_____

_____

Part 3 *Thinking*

1. If we want to obtain the striped fabrics? How many kinds of dyeing methods you can use?

_____

_____

_____

2. We usually find that clothes are easy to fade in the washing process, which indicates that the dyeing fastness of the fabric is not good? So would you please think about it or check the relevant information to explain what dyeing fastness is? What are the indicators for evaluating dyeing fastness?

Project Seven Textile Dyeing and Finishing(纺织品染整)

Task Three Finishing Technology(整理技术)

【Text】

纺织品整理是一个通用术语,通常指纺织品在染色或印花后、在投放市场之前的处理[1]。纺织品整理的一般目的是完善纺织品并使其适合最终用途[2]。因此,纺织品整理可以增加纺织产品的附加值。根据纺织品整理后获得的使用功能,可分舒适功能纺织品、防护功能纺织品、卫生保健功能纺织品,以及其他功能的纺织品。就技术特征而言,整理加工过程可以分成两类:机械整理与化学整理。机械(整理)工序需要一些将纺织材料通过机械作用达到预期效果的机器[3]。而且,机械整理通常伴随着加热过程以增强所需效果[4]。例如,为了改善织物的手感和外观,可以在织物上进行硬挺整理、柔软整理和轧光整理等机械整理[5]。

【Key Words】

textile finishing 纺织品整理
perfect (动词)使完善,使完美
render ['rendə] 使变为,使变得
added value 附加值
comfortable textiles 舒适功能纺织品
protective textiles 防护功能纺织品
hygiene and health textiles 卫生保健功能纺织品
technical feature 技术特征

mechanical finish 机械整理
chemical finish 化学整理
handle 手感
appearance 外观
accompany [ə'kʌmpəni] 伴随,陪伴
stiffening ['stifniŋ] 硬挺整理
softening ['sɔfniŋ] 柔软整理
calendering ['kælindəriŋ] 轧光整理

【Key Sentences】

1. **Textile finishing** is a general term which usually refers to treatments on textile fabrics after dyeing or printing but before they are put on the market.

Fig 7.23 shows classification of functional textiles.

2. The general aim of the textile finishing is to **perfect** the textile goods and **render** them fit for their end use.

3. Mechanical process involve machines that make textile materials achieve the desired effects through mechanical action.

4. Mechanical process frequently **accompanied** by a heating process to enhance the desired effects.

5. For example, to improve the **handle** and **appearance** of the fabrics, **stiffening**, **softening** or

calendering etc., could be exerted on the fabric.

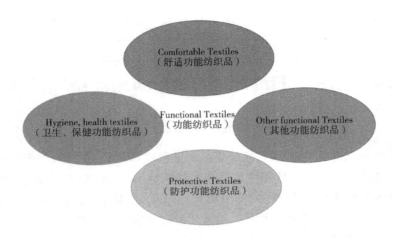

Fig 7.23 Classification of functional textiles(功能纺织品的分类)

【Text】

　　硬挺整理利用天然或合成的高分子材料制成具有一定黏度的浆液,在织物上形成薄膜,从而使织物获得平滑、硬挺、厚实、丰满的手感,并提高其强力和耐磨性,延长使用寿命[1]。硬挺整理广泛应用于箱包布、帐篷布和广告布等织物的后整理中。柔软整理有机械的和化学的两种方法。机械柔软整理利用机械方法,在张力状态下,将织物多次揉屈,以降低织物的刚性,使织物恢复至适当的柔软度[2];化学柔软整理利用柔软剂的作用来降低纤维间的摩擦系数以获得柔软效果[3]。柔软整理广泛应用于棉织物与麻织物。

　　轧光本质上是一个增加织物光泽的熨烫过程。在一定的机械压力、温度、湿度的作用下,借助纤维的可塑性,纱线被压扁,织物表面被压平,以提高织物表面光泽及光滑平整度[4]。轧光机是至少由两个轧辊组成的重型机器,通过更换轧辊的表面结构来获得不同的轧光效果;轧光整理一般可分为普通轧光(织物压平而获得光泽)、电光(织物表面压出平行而整齐的斜纹线,对入射光规则反射,获得如丝绸般的光泽)、轧纹(织物表面产生凹凸花纹,从而形成隐约的花纹效果)等[5]。轧光整理在棉针织物、针刺非织造织物过滤材料上的应用比较多。

【Key Words】

polymeric　聚合的
seriflux　浆液
plump [plʌmp]　丰满的
service life　使用寿命
luggage cloth　箱包布
tent cloth　帐篷布
advertising cloth = inkjet cloth　广告布喷绘布

knead [niːd]　揉屈
rigidity [rːˈdʒidəti]　刚度
softener [ˈsɒfnə(r)]　柔软剂
friction coefficient [ˌkəuiˈfiʃnt]　摩擦系数
plasticity [plæˈstisəti]　可塑性,塑性
flatten [ˈflætn]　使压扁
crush　使压平
calender　轧光机

ordinary calendering 普通轧光整理	pattern calendering 轧纹整理
electric calendering 电光整理	bump 隆起,凸块
incident light 入射光	vague [veig] 模糊的,不清楚的

【Key Sentences】

1. Stiffening: Use natural or synthetic **polymeric** material to make a certain viscosity of the **seriflux** which forms a film on the fabric, so that the fabric could achieve smooth, hard, thick, and **plump** feel, improve its strength and wear resistance, and the **service life** extend.

2. Mechanical softening: Use mechanical methods to make the fabric be **kneaded** several times under the tension and reduce the **rigidity**, so that the fabric can be restored to the appropriate softness.

3. Chemical softening: Use the action of **softener** to obtain a soft effect by reducing the **friction coefficient** between the fibers.

4. Under the influence of a certain mechanical pressure, temperature and humidity, due to the fiber's **plasticity**, the yarns are **flattened** and the surface of the fabric was **crushed** in order to improve the surface gloss and smooth ness of the fabric.

5. Calendering can generally be divided into **ordinary calendering** (The surface of the fabric is flattened to obtain luster), **electric calendering** (The surface of the fabric is pressed out of a parallel and neat twill line, which make the fabric reflects the **incident light** regularly and obtain the silky luster), **pattern calendering** (The surface of the fabric produces a **bump** pattern, resulting in a **vague** pattern effect) and so on.

【Text】

定形整理包括拉幅和机械预缩两种整理。拉幅是织物的机械矫直[1]:因纤维在潮湿状态下具有一定的可塑性能,可将织物幅宽缓缓拉宽至规定的尺寸,并符合印染成品的规格要求[2];常使用布铗或针板拉幅机,织物在布铗或针板之间被两边的布边水平握持;握持织物的链条在进入干燥装置之前逐渐将其分开到所需的宽度[3]。机械预缩整理是利用机械物理方法改变织物中经向纱线的屈曲状态,使经纱屈曲增加,织物的长度缩短且结构松弛,从而消除经向的潜在收缩[4]。

【Key Words】

stablishing 定形整理	specification [ˌspesifiˈkeiʃn] 规格
tentering 拉幅整理	clip tenter 布铗拉幅机
sanforizing 机械预缩整理	pin tenter 针板拉幅机
straighten [ˈstreitn] 使变直,使拉直	chain 链条
specified [ˈspesifaid] 规定的	apart 分开

【Key Sentences】

1. **Tentering** is the mechanical straightening of fabrics.

2. The fiber has a certain plasticity in wet state, so the fabric width can be slowly widened to the **specified** size and meet the **specifications** of the finished product.

3. The **chains** holding the fabric gradually move **apart** to the desired width before the cloth entering the drying unit.

4. **Sanforizing**: Use mechanical/physical methods to change the bending state of the warp yarns in the fabric, that is, it increases warp yarn's bending, so fabric length is shortened and the fabric structure becomes loose, thus eliminate the potential shrink of the fabric in warp direction.

【Text】

机械整理和化学整理之间的区别难以区分,因为在许多机械整理中使用化学品或在许多化学整理过程中纺织品受到机械作用[1]。也许,常见的整理过程应该称为组合整理。这意味着机械和化学整理在整理过程中都用于纺织品。化学整理可以描述为将化学品施加到织物上的加工过程[2]。化学整理剂通常以水溶液或乳液的形式运用,并且可以通过多种技术施用,主要技术是浸轧法[3]。

纺织品不仅能用于一般日常生活,经过一些特殊的整理加工后,还可以获得一些特殊应用,如拒水、阻燃、防静电、防污等[4]。一般纺织品并不具备这些性能,而是经过特殊的化学整理方法获得,这类整理方法称为特种整理[5]。随着人们对纺织品特殊功能的需求越来越多且整理技术突飞猛进,特种整理的种类变得越来越多,其发展也越来越快[6]。

【Key Words】

distinction [diˈstiŋkʃn] 差别,区别
combined finishing 组合整理
aqueous [ˈeikwiəs] 水的,含水的,水状的
emulsion [iˈmʌlʃn] 乳状液,乳浊液
pad mangle 浸轧

water-repellenting 拒水
flame retardant 阻燃
anti-static 抗静电
anti-fouling 防污
special finishing 特种整理

【Key Sentences】

1. The **distinction** between the mechanical and the chemical finishing often fails, as in many mechanical processes, chemicals are used or in many chemical processes the textiles are subjected by the mechanical action.

2. Chemical processes may be described as those processes which involve the application of chemicals to the fabric.

3. Chemical finishes are normally applied in the form of an **aqueous** solution or **emulsion** and may be applied via a variety of techniques, the main one being the **pad mangle**.

4. Textiles can not only be used in the daily life, it can also obtain some special applications after some special finishing processes, such as **water-repellenting**, **flame retardant**, **anti-static**, **anti-fouling** and so on.

5. These special properties are not available in conventional textiles, but they can be achieved by special chemical finishing methods which called **special finishings**.

6. With more and more demand for special functional textiles and the rapid development of finishing technology, there are more and more types of special finishings whose development get faster and faster.

【Text】

赋予织物吸收、阻断紫外线,保护皮肤不被伤害的功能整理工艺称为防紫外整理[1]。将紫外线吸收剂通过涂层或浸渍的方法整理到织物上,当紫外线吸收剂遇到紫外线后,它们能吸收紫外光,并将之转化为无害的辐射波,或将紫外线反射出去,从而起到保护作用[2]。防紫外整理效果与织物的组织(覆盖系数)和颜色密切相关,密而厚、颜色深的纺织品防紫外整理效果好。防紫外线整理主要应用于棉纤维与涤纶,以制作夏天服装为多,如运动服装、运动休闲装、防晒衣以及遮阳伞等[3]。

【Key Words】

ultraviolet [ˌʌltrəˈvaiələt] ray 防紫外线
anti-ultraviolet finishing 防紫外线整理
UV absorbers 紫外吸收剂,紫外线吸收剂,紫外光吸收剂
convert……into…… 把……转化为……
radiation wave 辐射波
covering coefficient [ˌkəuiˈfiʃnt] 覆盖系数
sports casual wear 运动休闲装
umbrella 伞

【Key Sentences】

1. The functional finishing process which gives the fabric absorption, blocking **ultraviolet rays** thus protecting the skin from injury is called **anti-ultraviolet finishing**.

2. The **UV absorbers** are coated or impregnated onto the fabric, so when ultraviolet absorbers encounter ultraviolet light, they can absorb UV rays and **convert** them **into** harmless **radiation waves**, or reflect ultraviolet rays, thus playing a protective role.

Fig 7.24 shows the principle of anti-ultraviolet textiles.

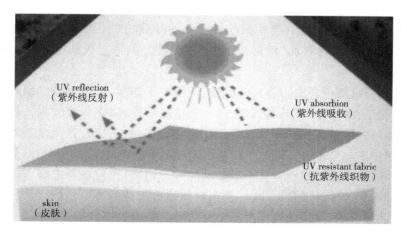

Fig 7.24　Principle of anti-ultraviolet textiles(防紫外线纺织品原理)

3. Anti-ultraviolet finishing is mainly used in cotton fiber and polyester fiber to make summer clothing, such as sportswear, **sports casual wear**, sunproof dothing, as well as **umbrellas**.

Fig 7.25 shows some anti-UV textiles.

(a) Anti-ultraviolet umbrella
（防紫外线遮阳伞）

(b) Sunproof clothing
（防晒衣）

(c) Uv resistant sportswear
（防紫外线运动套装）

(d) UV resistant casual wear
（防紫外线休闲服）

Fig 7.25 Anti-UV textiles（防紫外线纺织品）

【Text】

纤维、纱线或织物在加工或使用过程中由于摩擦而带静电。纤维带有静电后,容易吸附尘垢而造成沾污[1]。在加工过程中,由于电荷的吸引或排斥,会使(纺织)加工困难[2]。带有静电的衣服,则会贴附人体或互相缠附。当积聚的静电高于500V时,因放电而产生火花,会引起火灾[3]。将化学整理剂施于纤维表面,增加纤维表面亲水性,以防止在纤维上积聚静电的整理工艺,则为抗静电整理[4]。

防静电整理面料可用于开发防静电劳保服、防静电工作服、防静电洁净服等。防静电劳保服主要用于石油化工、油轮港口、兵器医药、油库等危险场所,防止服装或人体静电放电引起爆炸、火灾事故发生[5];防静电工作服主要用于电子通信设备加工,ESD产品制造与组装及计算机房等工作场所,保护产品免受破坏和避免仪器由于静电干扰引起错误动作[6];防静电洁净服主要用于电子、医药、食品、生物工程等具有洁净要求的生产环境,可避免静电吸尘或屏蔽人体产生微尘进入工作空间,影响产品质量[7]。这些应用对防静电服装的各个指标以及性能都有严格要求;此外,军队与武警常服、家用地毯也都需要有防静电功能。

Project Seven Textile Dyeing and Finishing(纺织品染整)

【Key Words】

electrostatic [iˌlektrəu'stætik] 静电的
static electricity 静电
contamination [kənˌtæmi'neiʃn] 混合,指搀合,合在一起
exclusion [ik'sklu:ʒn] 排斥,排除在外
charge 电荷
be attached to 黏附在
accumulate [ə'kju:mjəleit] 积累,积聚
volt [vəult] 伏,伏特
spark 火花
discharge 放电
hydrophilicity 亲水性
anti-static protective clothing 防静电劳保服
anti-static work clothing 防静工作服
anti-static clean clothing 防静电洁净服
petrochemical 石油化工
oil tanker ['tæŋkə(r)] (轮船)
port 油轮港口
oil depot ['depəu] 油库
explosion [ik'spləuʒn] 爆炸
fire accident 火灾
electrostatic [iˌlektrəu'stætik] 静电的
discharge 放电
electronic communication equipment 电子通信设备
static interference 静电干扰
electronics [iˌlek'trɔniks] 电子
biological [ˌbaiə'lɔdʒikl] engineering 生物工程
electrostatic precipitation [priˌsipi'teiʃn] 静电吸尘
shield [ʃi:ld] 屏蔽
military ['militri] 军队的
armed police 武装警察
uniforms ['ju:nifɔ:mz] 制服

【Key Sentences】

1. Fiber with **static electricity** is easy to absorb dust and cause **contamination**.

2. In the processing, due to the attraction or **exclusion** of **charges**, it will make processing difficult.

3. When the **accumulated** static electricity is higher than 500 **volts**, **sparks** due to **discharge** can cause a fire.

4. **Antistatic finishing** is the process of applying chemical finishing agent on the surface of the fibers to increase the **hydrophilicity** of the fibers so as to prevent the accumulation of static electricity on the fibers.

Fig 7.26 shows the principle of antistatic textiles.

5. **Anti-static protective clothing** is mainly used in **petrochemicals**, **oil tanker ports**, weapons, medicine, **oil depots** and other dangerous places, because it can prevent **explosions** and **fire accidents** caused by **electrostatic discharge** of clothing or human body.

Fig 7.27 shows some antistatic textiles.

6. **Anti-static work clothing** is mainly used in **electronic communication equipment** processing, ESD product manufacturing and assembling, computer room and other working places, because it can prevent products from damage and avoid the wrong action of instrument caused by **static interfer-**

Fig 7.26　Principle of antistatic textiles(抗静电纺织品原理)

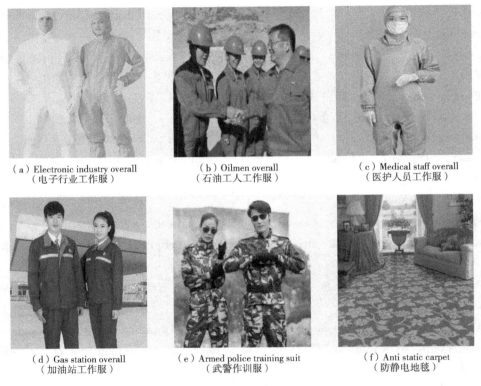

Fig 7.27　Antistatic textiles(抗静电纺织品)

ence.

7. **Anti-static clean clothing** is mainly used in **electronics**, medicine, food, **biological engineering** and other production environment with clean requirements, it can avoid **electrostatic precipitation** or **shielding** the human body to produce dust in to the work space which may affect the product quality.

Project Seven　Textile Dyeing and Finishing(纺织品染整)

【Text】

吸湿速干整理是针对涤纶、锦纶及其他化学纤维织物吸湿排汗性差的特点,对织物表面进行树脂涂层整理[1]。经过整理后,织物具有良好的吸汗性、毛细管透水透气性,可迅速将汗水吸尽并将其和湿气导离皮肤表面[2];同时,吸湿速干整理能克服织物燥身、不吸汗或潮湿衣物粘身、不易干等现象,使人们在夏季等高湿热环境下穿着具有清凉感[3]。吸湿速干纺织品主要应用在运动服、职业装、休闲服(T恤、衬衣、帽等)、内衣、袜子、毛巾等。

【Key Words】

moisture transferring and quick drying finishing　吸湿快干整理
resin ['rezin] 树脂(一种涂层整理剂)
perspire [pə'spaiə(r)] 出汗,排汗
perspiration [ˌpə:spə'reiʃn] 排汗
capillary [kə'piləri] 毛细血管
permeability [ˌpə:mjə'biliti] 通透
suck up 吸收

overcome　克服
phenomena [fi'nɔminə] 现象
humid　潮湿的
sportswear　运动服
professional wear　职业装
casual wear　休闲装
sark [sɑrk] 衬衣
underwear　内衣

【Key Sentences】

1. **Moisture transferring and quick drying finishing** is aimed at polyester, nylon and other chemical fiber fabrics which have poor moisture absorption and perspiring, to do **resin** coating finish on their surface.

Fig 7.28 shows the principle for moisture transferring and quick drying textiles.

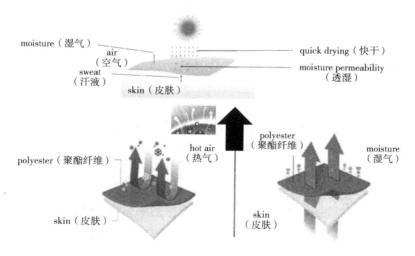

Fig 7.28　Principle for moisture transferring and quick drying textiles(吸湿速干纺织品原理)

2. After finishing the fabrics have a good **perspiration**, **capillary permeability**, so it can quick-

ly **suck up** sweat and moisture and remove them from the skin surface.

3. It can **overcome** the **phenomena** that the fabric does not suck **sweat**, damp clothing adhere to the skin and not easy to dry, and it also gives people a cool feel when they wear it in high hot and **humid** environments such as summer.

Fig 7.29 shows some moisture transferring and quick drying textiles.

Fig 7.29 Some moisture transferring and quick drying textiles(部分吸湿速干纺织品)

【Text】

红外线是在所有太阳光中最能够深入皮肤和皮下组织的一种射线[1]。远红外线与人体内细胞分子的振动频率接近,远红外线渗入体内之后,便会引起人体细胞原子和分子的共振吸收远红外线[2]。因此,远红外加速了血液的循环,有助于清除血管囤积物及体内有害物质,达到活化组织细胞、防止老化、强化免疫系统的目的[3]。总之,远红外线对于血液循环和微循环障碍引起的多种疾病均具有改善和防治作用[4]。远红外整理是采用高科技手段把远红外粉加工成15nm以下的颗粒胶体及其整理液,对棉、毛、涤纶、锦纶及其各类纺织品进行整理[5]。远红外整理后的纺织品具有高效的远红外反射能力和保温性能,也能改善人体血液循环、降低血液酸性、促进新陈代谢、促进肌体再生、增强免疫功能,延年益寿[6]。

基于远红外纺织品的性能特点,远红外织物适宜做防寒织物、轻薄型的冬季服装;适宜制作贴身内衣、袜子、床上用品,以及护膝、护肘、护腕等,远红外纺织品对金黄色葡萄球菌、白色念珠菌、大肠杆菌等病菌的抑菌率达98%,利用这些特性可制作卫生、医疗用品等产品。

【Key Words】

far-infrared ray 远红外光线
subcutaneous [ˌsʌbkjuˈteɪnɪəs] 皮下的
tissue [ˈtɪʃuː] 组织
resonance [ˈrezənəns] 共鸣,共振
atom 原子
molecule 分子
blood circulation 血液循环
vascular [ˈvæskjələ(r)] 血管的,脉管的
hoarding [ˈhɔːdɪŋ] 囤积
aging [ˈeɪdʒɪŋ] 变老,老化
immune system 免疫系统
disorder [dɪsˈɔːdə(r)] 失调,紊乱
blood circulation 血液循环
microcirculation 微循环
granular [ˈgrænjələ(r)] 由颗粒构成的,含颗粒的
colloidal [kəˈlɔɪdəl] 胶体的,胶质的,胶状的
heat insulating performance 保温性能

reflection [rɪˈflekʃn] (声、光、热等的)反射
acidity [əˈsɪdəti] 酸性
metabolism [məˈtæbəlɪzəm] 新陈代谢
body regeneration 肌体再生
immune [ɪˈmjuːn] 有免疫力的
promote longevity [lɔnˈdʒevəti] 延年益寿
cold-proof 防寒的
bedding 床上用品
pad [pæd] 软垫,护垫
knee pads 护膝
elbow pads 护肘
wrist pads 护腕
antiseptic [ˌæntɪˈseptɪk] rate 抑菌率
aureus staphylococcus [ˌstæfɪləˈkɔkəs] 金黄色葡萄球菌
albicanscandida [ˈkændɪdə] 白色念珠菌
escherichia coli 大肠杆菌

【Key Sentences】

1. **Infrared ray** is one of the most deep rays in all sunlight that can penetrate into the skin and **subcutaneous tissue**.

2. The vibration frequency of the far infrared ray is close to that of the cell molecules in the human body, when the infrared ray penetrates into the body, it will cause the **resonance** of the **atoms** and **molecules** in human cells which absorb the energy of the infrared ray.

Fig 7.30 shows the principle of far infrared textiles.

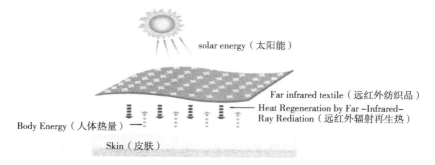

Fig 7.30 Principle of far infrared textiles(远红外纺织品原理)

3. As a result, the far infrared rays accelerate the **blood circulation**, help remove **vascular hoarding** and harmful substances in the body, and achieve the purpose of activating tissue cells, preventing **aging** and strengthening the **immune system**.

4. In a word, the far infrared ray has the effect of improving and preventing many kinds of diseases caused by the **disorder** of **blood circulation** and **microcirculation**.

5. Far-infrared finishing is the use of high-tech means to process far-infrared powder into less than 15 nm **granular colloidal** and its finishing solution, which is finished on cotton fabric, wool fabric, polyester, nylon and various textiles.

6. Far infrared finishing textiles have high efficiency of far-infrared **reflection** and **heat insulating performance**, but it also can improve human blood circulation, reduce blood **acidity**, promote **metabolism** and **body regeneration**, enhance **immune** function, and **promote longevity**.

Fig 7.31 shows some far infrared textiles.

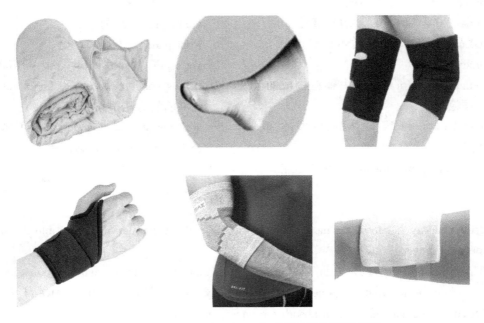

Fig 7.31　Some far infrared textiles(远红外纺织品原理)

【Text】

许多芳香剂具有镇静、杀菌、催眠、保健等作用,常规的芳香整理是将芳香剂简单地喷洒在纺织品上,但留香期很短[1]。目前采用最多的芳香整理技术用芳香剂作囊芯制备芳香微胶囊,并将芳香微胶囊整理到织物上[2]。芳香剂微胶囊化后,芯材与外界环境隔绝,但性质基本不变[3];开始时,在微胶囊外层的香精散发香味,在穿着过程中,由于摩擦、受热等外力作用,微胶囊内部的香精缓缓地释放香味[4];因此,微胶囊起到长效缓释的作用,从而使纺织品具有相当持久的芳香功能[5]。因此,芳香整理满足了人们对芳香纺织品的需求。芳香整理主要用于一些日常生活用品,如毛巾、窗帘、床上用品、地毯、桌布等[6]。

Project Seven Textile Dyeing and Finishing(纺织品染整)

【Key Words】

aromatic [ˌærəˈmætik] 芳香的,有香味的
aromatic agent 芳香剂
sedation [siˈdeiʃn] 镇静,镇定
sterilization [ˌsterəlaiˈzeiʃ(ə)n] 灭菌
hypnosis [hipˈnəusis] 催眠
health care 保健
fragrance [ˈfreigrəns] 芳香,香味
fragrance period 留香期
perfume [ˈpəːfjuːm] 芳香,香味
capsule core 囊芯

microcapsule 微胶囊
external [ikˈstəːnl] 外部的
isolate [ˈaisəleit] (使)隔离,孤立
emit [iˈmit] 发出,射出,散发
long-term 长期的
slow-release 慢释,缓释
daily necessities 生活必需品
towel [ˈtauəl] 毛巾,手巾
tablecloth 桌布

【Key Sentences】

1. Many **aromatic** agents have **sedation**, **sterilization**, **hypnosis**, **health care** and other functions, and regular aromatic finishing just spray aromatic agents on textiles simply, but the **fragrance period** is very short.

2. At present, the most used aromatic finishing technology uses the aromatic agent as a **capsule core** and prepares the aromatic **microcapsules** which are finished onto the fabric.

3. After the microcapsule of the aromatic agent, the core material and the **external** environment are **isolate**d, but its nature is basically unchanged.

4. At the beginning, the flavor outside of the microcapsules **emits** perfume, in the process of wearing, the flavor inside the microcapsules slowly releases the fragrance due to friction, heat and other external forces.

Fig 7.32 shows the principle for perfume textiles.

Fig 7.32 Principle for perfume textiles(芳香纺织品原理)

5. So the microcapsules have the **long-term** and **slow-release** effect which make the textiles a fairly lasting aromatic function.

6. Aromatic finishing is mainly used in some **daily necessities**, such as towels, curtains, **bedding**, carpets, **tablecloth** and so on.

243

Fig 7.33 shows some perfume textiles.

(a) Aromatic towel
（芳香毛巾）

(b) Aromatic sheets
（芳香床单）

(c) Aromatic curtain
（芳香窗帘）

(d) Aromatic carpet
（芳香地毯）

(e) Aromatic table cloth
（芳香桌布）

(f) Aromatic ladies' underwear
（芳香女士内衣）

Fig 7.33　Perfume textiles（芳香纺织品）

【Text】

　　季铵表面活性剂和抗菌素可起到抗菌剂的作用；许多含氯有机化合物和含有铜、银、铁、锰或锌的有机金属化合物也能使纺织材料抵抗微生物的生长[1]；含氯有机化合物也能有效地抑制昆虫对由蛋白质和纤维素组成的天然纤维的攻击[2]。抗菌整理是用上述提到的一些抗菌剂处理织物，从而获得抗菌、防霉、防臭、保持纺织品清洁卫生的加工工艺[3]。抗菌整理赋予纺织品一种特殊的功能，防止织物被微生物沾污而损伤[4]；更重要的是抗菌整理能防止疾病传染，保证人体的安全健康，降低公共环境的交叉感染率，使织物获得卫生保健的新功能[5]。

　　抗菌整理主要应用于被大部分人连续接触的纺织品[6]。抗菌整理纺织品广泛用于人们的内衣、睡衣、运动衣、袜子、鞋衬布、婴儿尿布、床垫布；也能用于医院、宾馆、机构住房的床单、被套、枕头、毛毯、餐巾、毛巾、沙发布、窗帘布、地毯、家具装饰布；此外，医药、食品、服务行业的工作服，部队服装以及绷带、纱布上也采用抗菌整理[7]。

【Key Words】

quaternary ammonium surfactants　季铵盐表面活性剂

antibioctics [ˌæntibaiˈɔtiks] 抗菌素

chlorinated organic compounds　含氯有机化合物

organometallic　有机金属的

copper　铜

silver　银

iron　铁

manganese [ˈmæŋɡəniːz] 锰

zinc 锌
microorganism [ˌmaikrəuˈɔːgənizəm] 微生物
chlorine [ˈklɔːriːn] 氯
anti-mold 防霉
anti-odor [ˈænti ˈəudə] 防臭
hygienic [haiˈdʒiːnik] 卫生的
microbial [maiˈkrəubiəl] 微生物的，细菌的
contamination [kənˌtæmiˈneiʃn] 污染
well-being 健康，康乐
cross-infection 交叉感染

pajamas [pəˈdʒɑːməz] 睡衣
sportswear 运动衣
shoe lining 鞋衬
baby diaper 婴儿纸尿裤
mattress ticking 床垫布
institutional homes 机构住房
bed sheet 床单
quilt [kwilt] cover 被套
pillow 枕头
upholstery [ʌpˈhəulstəri] 家具装饰布
bandage [ˈbændidʒ] 绷带

【Key Sentences】

1. Many organic compounds containing chlorine and **organometallic** compounds containing **copper**, **silver**, **iron**, **manganese**, or **zinc** also make textile materials resist growth of **microorganisms**.

2. The organic compounds containing **chlorine** are also effective in inhibiting attack of insects on natural fibers composed of protein and cellulose.

3. Antibacterial finishing is a kind of processing technology which use some of the above mentioned antibacterial agent to treat the fabric, so as to obtain the effect of antibacterial, **anti-mold**, **anti-odor**, and keep textile clean and **hygienic**.

Fig 7.34 shows the principle of antibacterial textiles.

4. It gives textiles a special function to prevent fabrics from being damaged by **microbial contamination**.

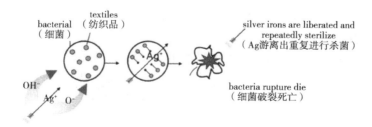

Fig 7.34 Principle of antibacterial textiles(抗菌纺织品原理)

5. What's more, it prevents diseases, ensures the safety and **well-being** of the human body, reduces the **cross-infection** rate of the public environment, and enables fabrics to gain new features in health care.

6. The main use of antibacterial finishing is for textiles which are being handled continuous by a large number of people.

7. In addition, work clothes in medicine, food and service industry, troop clothing and **bandages**, **gauze** also be treated by antibacterial finishing.

Fig 7.35 shows some antibacterial textiles.

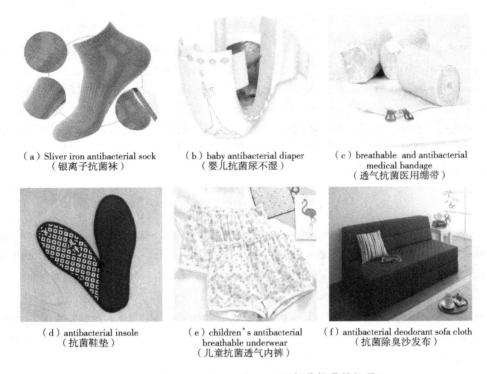

Fig 7.35 Some antibacterial textiles(部分抗菌纺织品)

【Text】

  一般情况下,纺织品的燃烧基本都是外部热源作用于自身[1]。当热源提供的热量到达一定程度后,纺织品中的纤维就会发生分解或者裂解,并产生一定量的可燃气体[2];可燃气体与氧气混合后,纺织品发生燃烧现象[3]。阻燃整理用无机或有机阻燃剂对纤维或织物进行表面涂层[4]。大多数纺织阻燃剂会改变或中断聚合物正常热分解过程,可减少热分解过程中可燃性气体的生成和阻碍气相燃烧过程中的基本反应,或吸收燃烧区域中的热量[5];此外,稀释和隔离空气对阻止燃烧也有一定的作用[6]。

  许多纺织纤维,特别是纤维素纤维的局限性(缺点)是它们容易燃烧。而阻燃整理的目的是赋予纺织品阻燃性[7]。阻燃纺织品主要用于消防员工作服、钢铁厂工人工作服、军用纺织品、悬挂窗帘、地毯、童装等特殊领域。

【Key Words】

combustion [kəmˈbʌstʃən] 燃烧　　　　crack 使裂解

decompose [ˌdiːkəmˈpəuz] 使分解　　　flammable [ˈflæməbl] 易燃的,可燃的

oxygen ［'ɔksidʒən］氧气
flame retardant ［ritɑ:dənt］finishing 阻燃整理
inorganic ［,inɔ:'gænik］无机的
decomposition ［,di:,kɔmpə'ziʃn］分解（名词）
generation ［,dʒenə'reiʃn］产生
gas phase 气相
dilute ［dai'lu:t］稀释（动词）
burn ［bə:n］燃烧
readily ［'redili］轻而易举地

flame retardant ［ritɑ:dənt］finish 阻燃整理
impart ［im'pɑ:t］ 把（某性质）赋予，将…给予
flame resistance 阻燃性
firemen 消防队员
steel plant 炼钢厂,钢铁厂
military ［'milətri］军事的,军队的
hanging curtain 悬挂窗帘
children's wear/clothing 童装

【Key Sentences】

1. In general, the **combustion** of textiles is basically caused by external heat source.
2. When the heat source provides a certain degree of heat, the fibers in textiles will be first **decomposed** or **cracked** and produce a certain amount of **flammable** gas.
3. After the flammable gas mixed with **oxygen**, then the textile will be burning.

Fig 7.36 shows the diagram of the textile combustion process.

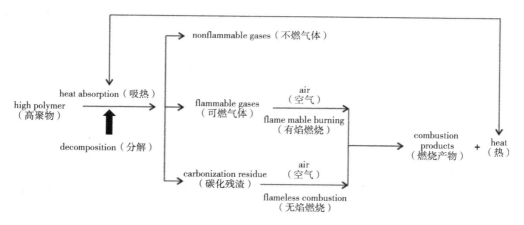

Fig 7.36 Diagram of the textile combustion process(纺织品燃烧过程示意图)

4. **Flame retardant finishing**: Make **inorganic** or organic flame retardants coating on the surface of fibers or fabrics.
5. Most textile flame retardants will change or interrupt the normal thermal **decomposition** process of polymers, so they reduce the **generation** of flammable gases during thermal decomposition process and hinder the basic reaction in the combustion process of **gas phase**, or absorb heat from the combustion area.
6. In addition, **diluting** and isolating air also play a role in preventing combustion.
7. The aim of **flame retardant finishing** is to **impart flame resistance** to textiles.

8. The flame-retardant textiles are mainly used in some special fields, such as working clothes of **firemen**, working clothes of **steel plant** worker, **military** textiles, **hanging curtain**, carpet, children's clothes, etc..

Fig 7.37 shows some flame-retardant textiles.

(a) Firefighter overall
(消防人员工作服)

(b) Iron workers' overall
(钢铁工人工作服)

(c) Flame-retardant carpets for hotel
(酒店用阻燃地毯)

(d) Flame-retardant children's wear
(阻燃童装)

(e) Flame-retardant curtain
(阻燃悬挂窗帘)

(f) Flame-retardant camouflage uniforme
(阻燃迷彩服)

Fig 7.37　Some flame-retardant textiles(阻燃纺织品)

【Text】

纤维素纤维制成的织物如纯棉、黏胶及其混纺织物具有很多优良的特性,但也存在着弹性差、易变形、易折皱等缺点,故在穿着过程中不能保持平整的外观[1]。为了改善上述不足之处,纤维素纤维织物可采用 N-羟基甲基类树脂、无甲醛类树脂整理剂等树脂(特殊的高分子预聚体)进行整理[2];(整理后)树脂在纤维素纤维内部通过交联作用、沉积作用增加了纤维的弹性,提高了织物从形变中回复的能力,防皱性能得到提高[3]。这个加工过程称为防皱整理,也可称为洗可穿整理、免烫整理、耐久轧烫整理(简称 PP 整理)。耐久轧烫整理是一种较新的防皱整理工艺,其先用树脂整理剂浸轧整理织物,使加工织物具有潜在的"热塑"性能;当织物制成衣片后,再进行高温压烫定形,使衣服具有耐久的折裥和稳定的外形。

纤维素纤维织物经防皱整理后,他们的力学性能如折皱回复角提高,而拉伸断裂强度、拉伸断裂延伸度、撕破强度和耐磨性方面也有明显的变化[4]。如防皱整理后,棉织物的断裂强度和断裂延伸度明显降低,降低的程度随织物防皱性能的提高而加剧[5];黏胶纤维经过防皱整理后由于在分子间建立了共价交联,使大分子间的作用力得到加强,纤维强度提高。但无论棉纤维还是黏胶纤维织物,经防皱整理后,拉伸断裂延伸度和撕破强度都发生明显的降低。此外,纤维素纤维织物经防皱整理后,织物的耐磨性随防皱性的提高而下降。因此,如果对纤维素纤维织物进行防皱整理,需要考虑这些指标的平衡。

天然蛋白质纤维如蚕丝和羊毛织物的弹性,虽然比纤维素纤维织物优良很多,但是与合成纤维织物相比,有一定的差距[6]。因此,近年来,对真丝织物的免烫整理和羊毛织物的防皱整理,进行了较多的研究,如对羊毛织物进行防毡缩整理,采用特殊的丝素整理剂用于真丝织物的抗皱整理[7]。

【Key Words】

poor elasticity　弹性差
easy deformation　易变形
easy wrinkling　易起皱
deficiency [di'fiʃnsi]　不足,缺点,缺陷
N-hydroxymethyl resin　N-羟基甲基类树脂
formaldehyde-free resin [fɔː'mældihaid]　无甲醛类树脂
macromolecule　高分子,大分子
prepolymer　预聚物
cross-linking　交联
deformation　变形
wrinkle resistance　抗皱性
anti-crease finishing　抗皱
wash-and-wear finishing　洗可穿整理
non-iron finishing　免烫整理
durable press finishing　耐久轧烫整理

pad　浸轧
potential　潜在的
thermoplastic　热塑性
folding　折裥
wrinkle recovery angle　折皱回复角
acting force　作用力
tensile breaking strength　拉伸断裂强度
tensile elongation at break　拉伸断裂延伸度
tear strength　撕破强度
abrasion resistance　耐磨性
aggravate ['ægrəveit]　使严重,使加剧
anti-crease performance　抗皱性能
covalent [ˌkəu'veilənt]　共价的
gap　差距
anti-felt finishing　防毡缩整理
fibroin finishing agent　丝素整理剂

【Key Sentences】

1. Fabrics made of cellulose fibers such as cotton, viscose and their blended textiles have many excellent properties, but there are also shortcomings such as **poor elasticity**, **easy deformation**, and **easy wrinkling**, so they can not maintain a flat appearance in the wearing process.

Fig 7.38 shows the wrinkled cotton fabric.

2. In order to improve the above-mentioned **deficiencies**, the cellulose fiber fabric may be finished by the resin such as an **N-hydroxymethyl resin** or a **formaldehyde-free resin** finishing agent (a special **macromolecule prepolymer**).

3. The resin increases the elasticity of the fiber through **cross-linking** and **deposition** inside the cellulose fiber, improves the ability of the fabric to recover from **deformation**, and improves the **wrinkle resistance**.

4. After thecellulose fiber fabrics are treated by anti-crease finishing, their physical and me-

Fig 7.38　Wrinkled cotton fabric（起皱后的棉织物）

chanical properties such as **wrinkle recovery angle** are improved, and **tensile breaking strength**, **tensile elongation at break**, **tear strength** and **abrasion resistance** are also significantly changed.

5. The tensile strength and elongation at break of the cotton fabric after being treated by anti-crease finishing are significantly reduced, and the degree of reduction is **aggravated** with the improvement of the **anti-crease performance** of the fabric.

6. Although the elasticity of natural protein fibers such as silk and wool fabrics is much better than cellulose fiber fabrics, it has a certain **gap** compared with synthetic fiber fabrics.

7. Therefore, in recent years, more research has been done on the non-iron finishing of silk fabrics and the anti-crease finishing of wool fabrics, such as **anti-felt finishing** of wool fabrics, and special silk **fibroin finishing agents** for anti-crease finishing of silk fabrics.

【Text】

在织物表面施加一种具有特殊结构的整理剂，使其牢固地附着于纤维或与纤维产生化学结合，从而改变了纤维表面层的组成，使织物不能被水和常用的食用油类所润湿，称为拒水拒油整理[1]。拒水拒油整理的实质是在织物表面施加一层氟碳化合物等特殊结构的物质，使其高能表面变为低能表面，以此获得拒水拒油效果的织物[2]。织物的表面能越小，拒水拒油效果越好。通过控制表面粗糙度与降低表面能，使与水或油在织物表面的接触角高于120°来达到拒水拒油效果[3]。

拒水拒油服装既能抵御雨水、油迹、寒风的入侵，保护肌体，又能让人体的汗液、汗气及时地排出，从而使人体保持干爽和温暖[4]。拒水拒油服装可以用于劳动保护服装、军服、运动服，以及钻井、测井等接触油水介质频繁的工人工作服，而且也可以用于制作风衣、雨衣[5]。此外，应用在装饰、产业领域中的具有拒水拒油功能的餐桌布、汽车防护罩、旅行包、帐篷等也备受青睐[6]。

【Key Words】

bound　（bind 的过去分词）结合
edible ['edəbl] 适宜食用的
essential difference　本质区别
water-repellenting　（名词）拒水

deposition [ˌdepə'ziʃn] 沉积
impermeable [im'pəːmiəbl] 不透水的，不透气的
filling　（名词）填充

waterproof 防水的
airproof 不透气的
essence ['esns] 本质,实质
water and oil repellent finishing 拒水拒油整理
fluorocarbon [ˌflurou'kɑrbən] 碳氟化合物
high-energy 高能
low-energy 低能
roughness [rʌfnəs] 粗糙
contact angle 接触角
invasion [in'veiʒn] 侵犯；干预
sweat 汗,出汗,流汗
sweat gas 汗气
discharge 排出,释放
labor protection clothing 劳动防护服
military clothing 军用服装
drilling ['driliŋ] 钻井
logging ['lɔgiŋ] 测井
trench coat 风衣
raincoat 雨衣
car protective cover 汽车防护罩
tent [tent] 帐篷
decoration [ˌdekə'reiʃn] 装饰
industrial field 产业用领域

【Key Sentences】

1. Applying a special structure finishing agent on the fabric surface, which is firmly attached or chemically **bound** to the fiber, thus it changes the composition of the fiber surface layer and the fabric can not be wetted by water and commonly used **edible** oil. The process is called **water and oil repellent finishing**.

2. The **essence** of **water-repellent and oil-repellent finishing** is to apply a special structure material such as **fluorocarbon** on the surface of the fabric to change its **high-energy** surface into **low-energy** surface, so as to obtain the fabric with water-repellent and oil-repellent effect.

3. By controlling the surface **roughness** and reducing the surface energy, the **contact angle** with water or oil on the surface of the fabric is higher than 120 degrees.

Fig 7.39 shows the principle for water and oil repellent textiles.

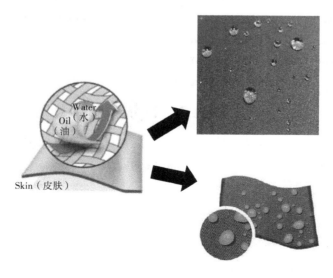

Fig 7.39 Principle for water and oil repellent textiles(防水防油纺织品原理)

4. Water-repellent and oil-repellent clothing can not only resist the **invasion** of rain, oil or cold wind, and protect the body, but also allow the **sweat** and **sweat gas** of the human body to be **discharged** in time, so as to keep the human body dry and warm.

Fig 7.40 shows some water and oil repellent textiles.

(a) Children's raincoat
（儿童雨衣）

(b) Adult poncho
（成人雨披）

(c) Outdoor jacket
（冲锋衣）

(d) Water and oil repellent table cloth
（拒水拒油餐桌布）

(e) Car cover
（汽车防护罩）

(f) Tent（帐篷）

Fig 7.40　Water and oil repellent textiles（拒水拒油纺织品）

5. Water-repellent and oil-repellent clothing can be used in **labor protection clothing**, **military clothing**, sportswear, **drilling** and **logging** workers' clothing which contacts oil and water frequently, and also can be used to make **trench coat**s and **raincoat**s.

6. Tablecloths, **car protective covers**, travel bags and **tents** with water and oil repellent functions are also popular in **decoration** and **industrial fields**.

Part 2　*Trying*

1. In addition to the chemical functional finishing in the textbook, have you seen or heard of any other functional finishing? What is its principle and function?

_____
_____
_____

2. Take stiffening finishing, softening finishing, far-infrared finishing, antibacterial finishing as

examples, write down 2~3 specific textiles respectively.

Finishing	Stiffening finishing	Softening finishing
Textiles		
	Far-infrared finishing	Antibacterial finishing
Textiles		

Part 3 *Thinking*

1. Combined with the entire process from fiber to functional textiles, taking the process of an antistatic medical worker's overalls as example, in order to make the overalls have anti-static functions, what methods can be obtained?

2. Considering the added value, what opportunities can the development of functional textiles bring to textile dyeing and finishing enterprises in the transformation and upgrading?

参考文献

[1] 孙钰,毛雷. 纺织专业英语[M]. 北京:中国劳动社会保障出版社,2010.
[2] 顾平. 纺织导论[M]. 北京:中国纺织出版社,2008.
[3] 张荣华. 纺织导论[M]. 上海:学林出版社,2012.
[4] 瞿才新. 纺织检测技术[M]. 北京:中国纺织出版社,2011.
[5] 刘华. 机织物分析与设计[M]. 上海:学林出版社,2012.
[6] 王成. 实用纺织英语[M]. 北京:中国纺织出版社,2013.
[7] 卓乃坚. 纺织英语[M]. 3版. 上海:东华大学出版社,2017.
[8] 蒋耀兴. 纺织概论[M]. 北京:中国纺织出版社,2005.
[9] 史志陶. 棉纺工程[M]. 北京:中国纺织出版社,2007.
[10] 崔鸿钧. 现代机织技术[M]. 上海:东华大学出版社,2010.
[11] 贺庆玉. 针织概论[M]. 北京:中国纺织出版社,2012.
[12] 言宏元. 非织造工艺学[M]. 北京:中国纺织出版社,2015.
[13] 范雪荣. 纺织品染整工艺学[M]. 北京:中国纺织出版社,2017.
[14] 张玉惕. 产业用纺织品[M]. 北京:中国纺织出版社,2009.
[15] 郁崇文. 纺纱学[M]. 3版. 北京:中国纺织出版社有限公司,2019.

(a) Cushion（坐垫）

(b) Bath towel（浴巾）

(c) Wall cloth（墙布）

(d) Dining chair/table cloth
（餐椅布、台布）

(e) Bed cover/pillowcase/curtain
（床罩、枕套、窗帘）

(f) Suit dust cover
（西服防尘罩）

(g) TV dust cover
（电视防尘罩）

(h) Air conditioning dust cover
（空调防尘罩）

(i) Non-woven bag
（非织造布袋）

(j) Blanket（毛毯）

(k) Lampshade cloth（布艺灯罩）

(l) Artificial grass（人造草皮）

Fig 1.2　Some decorative textiles(部分装饰用纺织品)

(a) Surgical gown, surgical cap,
mask, medical suture
（手术衣、手术帽、口罩、
医用缝合线）

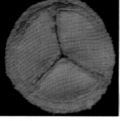
(b) Artificial heart valve
（人造心脏瓣膜）

(c) Diaper（尿不湿）

(d) Waist belt（护腰）

Fig 1.4　Some medical care and hygiene textiles(医疗卫生用纺织品)

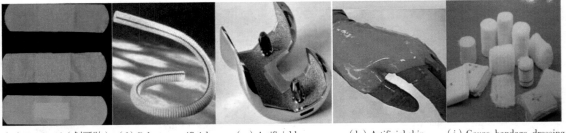

(e) Band-aid(创可贴)　　(f) Polyester artificial vessel （涤纶人造血管）　　(g) Artificial bone （人造骨）　　(h) Artificial skin （人造皮肤）　　(i) Gauze, bandage, dressing cloth（纱布、绷带、包扎布）

Fig 1.4　Some medical care and hygiene textiles(医疗卫生用纺织品)

(a) Water filter materials（水过滤材料）　　(b) Air filter materials （空气过滤材料）

(c) Anti-haze filter mask（防雾霾过滤口罩）　　(d) Fuel, oil filter and high temperature air filter materials （燃油、机油过滤材料及高温空气过滤材料）

Fig 1.5　Some filtration and separation textiles(部分过滤与分离用纺织品)

(a) Airbag（安全气囊）　　(b) Car seat cusion （汽车座椅）　　(c) Tire cord fabric （轮胎帘子布）

Fig 1.7　Some transport textiles(交通工具用纺织品)

(d) Seat belt（安全带）   (e) Car carpet（汽车地毯）   (f) Car curtain（汽车窗帘）

Fig 1.7   Some transport textiles（交通工具用纺织品）

(a) Pole vault rod（撑杆跳杆）   (b) Fishing pole（钓鱼竿）   (c) Golf clubs（高尔夫球杆）

(d) Wind turbine blades（风力发电叶片）   (e) Yacht hull（游艇船身）   (f) Aircraft door（飞机舱门）

Fig 1.9   Some structural reinforcement textile composites（结构增强纺织复合材料）

(a) Large stadium ceiling substrate   (b) Membrane materials for building（建筑用膜材料）
（大型体育场顶棚基材）

Fig 1.10   Some building and construction textiles（建筑用纺织品）

（c）Waterproof materials for construction
（建筑用防水材料）

（d）Architectural soundproof felt
（建筑用隔音毛毡）

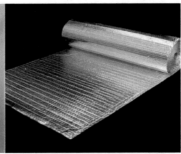
（e）Thermal insulation material for building
（建筑用隔热材料）

Fig 1.10　Some building and construction textiles（建筑用纺织品）

Fig 2.4　Samples of colored cotton（彩棉样品）

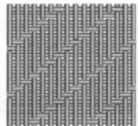

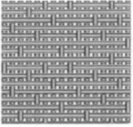

（a）Plain weave
（平纹）　　　　　　　（b）Twill weave
（斜纹）　　　　　　　（c）Satin weave
（缎纹）

Fig 4.33　The three basic weaves（三原组织）

（a）Socks（袜子）　　　　　　　　　　　　　　　（b）Underwear（内衣）

（c）Sweater（毛衣）　　　　　　（d）Trousers（长裤）　　　　　　（e）Mosquito net（蚊帐）

Fig 5.1　Knitting products（针织产品）

(f) Gloves（手套）　　(g) Suit（西服）　　(h) Carpet（地毯）　　(i) Scarf（围巾）

Fig 5.1　Knitting products（针织产品）

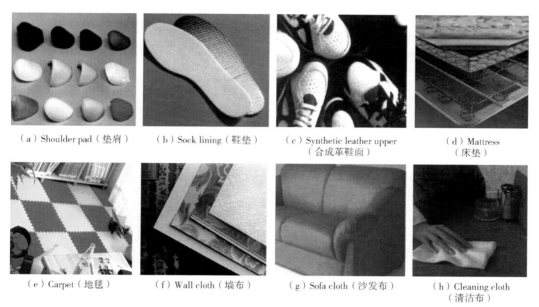

(a) Shoulder pad（垫肩）　　(b) Sock lining（鞋垫）　　(c) Synthetic leather upper（合成革鞋面）　　(d) Mattress（床垫）

(e) Carpet（地毯）　　(f) Wall cloth（墙布）　　(g) Sofa cloth（沙发布）　　(h) Cleaning cloth（清洁布）

Fig 6.4　Nonwoven fabrics used in clothing and shoes and household nonwoven fabrics
（用于服装、鞋及家居用非织造物）

(a) Packing bags（包装袋）　　(b) Thermal insulation film（保温膜）　　(c) Tent（帐篷）　　(d) Awning cover（遮阳蓬盖）

(e) Insulation materials（绝缘材料）　　(f) Filtration materials（过滤材料）　　(g) Trademark（商标）　　(h) Sorbent mat（吸油毡）

Fig 6.6　Nonwoven fabrics for industry and agriculture（工业用和农用非织造物）

Fig 7.6 Different colors of dying fabrics(不同颜色的染色织物)

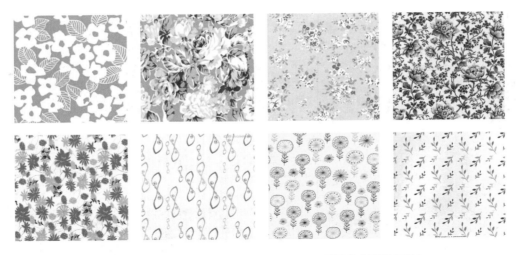

Fig 7.7 Different printing patterns of fabrics(不同印花图案织物)

(a) Natural environment-friendly dye (天然环保染料)　　(b) Dye solution(染液)　　(c) Reactive dyes(活性染色染料)

Fig 7.8 Preparation of water-soluable dye solution(可溶性染液的制备)

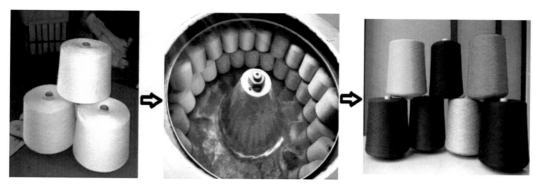

Fig 7.11　Bobbin yarn dyeing(筒纱染色)

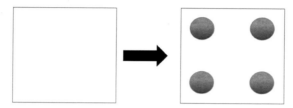

Fig 7.15　Diagram of direct printing(直接印花原理图)

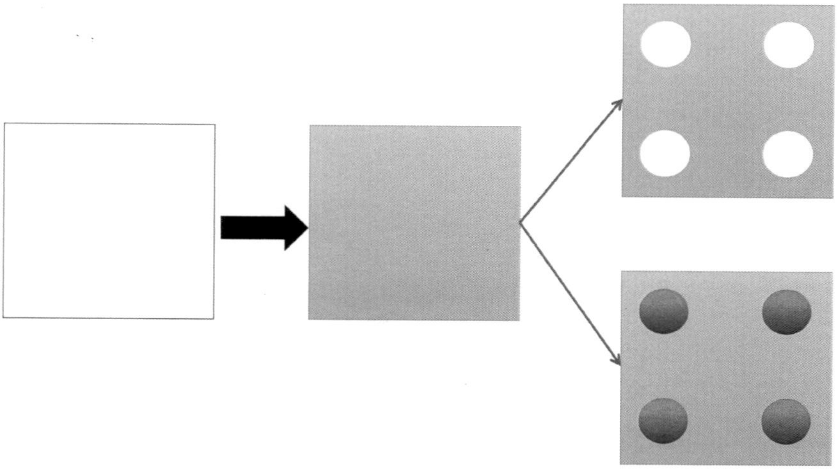

white discharge printing（拔白印花）

colour discharge printing（色拔印花）

Fig 7.16　Diagram of discharge printing(拔染印花原理图)

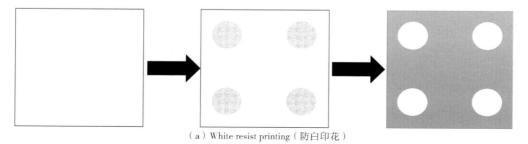

（a）White resist printing（防白印花）

Fig 7.17　Diagram of resist printing(防染印花原理图)

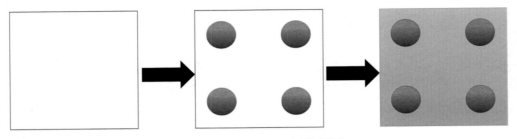

(b) Colour resist printing(色防印花)

Fig 7.17 Diagram of resist printing(防染印花原理图)

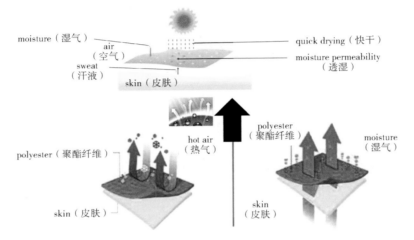

Fig 7.28 Principle for moisture transferring and quick drying textiles(吸湿速干纺织品原理)

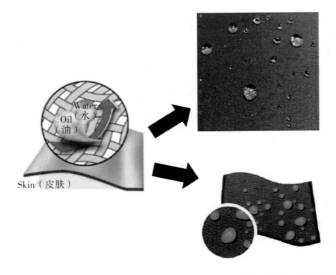

Fig 7.39 Principle for water and oil repellent textiles(防水防油纺织品原理)